50

MRE

Materials Research and
Engineering
Edited by B. Ilschner, N.J. Grant,
and Kenneth C. Russell

Krishan K. Chawla

Composite Materials
Science and Engineering

With 195 Figures and 37 Tables

Springer-Verlag

New York Berlin Heidelberg London Paris
Tokyo Hong Kong Barcelona Budapest

Prof. KRISHAN KUMAR CHAWLA

Dept. of Materials and Metallurgical Engineering
New Mexico Institute of Mining and Technology
Socorro, NM 87801/USA

Series Editors

Prof. BERNHARD ILSCHNER

Laboratoire de Métallurgie Mécanique
Département des Matériaux, Ecole Polytechnique Fedérale
CH-1007 Lausanne/Switzerland

Prof. NICHOLAS J. GRANT

Department of Materials Science and Engineering
Massachusetts Institute of Techology
Cambridge, MA 02139/USA

KENNETH C. RUSSELL

Department of Materials Science and Engineering and
Department of Nuclear Engineering, Massachusetts Institute of Technology,
Cambridge, MA 02139/USA

Library of Congress Cataloging-in-Publication Data
Chawla, Krishan Kumar, 1942–
Composite materials. (Materials research and engineering)
Includes bibliographies and indexes.
1. Composite materials. I. Title. II. Series.
TA418.9C6C43 1987   620.1'18   87-23379
ISBN 0-387-96478-9

Printed on acid-free paper.

Typeset by Asco Trade Typesetting Ltd., Hong Kong.
Printed and bound by Edwards Brothers, Inc., Ann Arbor, MI.
Printed in the United States of America.

9 8 7 6 5 4 corrected printing

ISBN 0-387-96478-9 Springer-Verlag New York Berlin Heidelberg
ISBN 3-540-96478-9 Springer-Verlag Berlin Heidelberg New York

आ नो भद्राः ऋतवो यन्तु विश्वतः

Ā no bhadrāḥ kratavo yantu viśvataḥ

Let noble thoughts come to us from every side

Rigveda 1-89-i

Dedicated affectionately

to Nivi and Nikhil and Kanika

Beechcraft Starship 1: An almost all composite aircraft
(Courtesy of Beech Aircraft Corporation)

Preface

The subject of composite materials is truly an inter- and multidisciplinary one. People working in fields such as metallurgy and materials science and engineering, chemistry and chemical engineering, solid mechanics, and fracture mechanics have made important contributions to the field of composite materials. It would be an impossible task to cover the subject from all these viewpoints. Instead, we shall restrict ourselves in this book to the objective of obtaining an understanding of composite properties (e.g., mechanical, physical, and thermal) as controlled by their structure at micro- and macrolevels. This involves a knowledge of the properties of the individual constituents that form the composite system, the role of interface between the components, the consequences of joining together, say, a fiber and matrix material to form a unit composite ply, and the consequences of joining together these unit composites or plies to form a macrocomposite, a macroscopic engineering component as per some optimum engineering specifications. Time and again, we shall be emphasizing this main theme, that is structure–property correlations at various levels that help us to understand the behavior of composites.

In Part I, after an introduction (Chap. 1), fabrication and properties of the various types of reinforcement are described with a special emphasis on microstructure–property correlations (Chap. 2). This is followed by a chapter (Chap. 3) on the three main types of matrix materials, namely, polymers, metals, and ceramics. It is becoming increasingly evident that the role of the matrix is not just that of a binding medium for the fibers but it can contribute decisively toward the composite performance. This is followed by a general description of the interface in composites (Chap. 4). In Part II a detailed description is given of some of the important types of composites, namely, polymer matrix composites (Chap. 5), metal matrix composites (Chap. 6), ceramic composites (Chap. 7), carbon fiber composites (Chap. 8), and multifilamentary superconducting composites (Chap. 9). The last two are described separately because they are the most advanced fiber composite systems of the 1960s and 1970s. Specific characteristics and applications of these composite systems are brought out in these chapters. Finally, in Part III, the micromechanics (Chap. 10) and macromechanics (Chap. 11) of composites are described in detail, again emphasizing the theme of how structure (micro and macro) controls the properties. This is followed by a description of strength and fracture modes in composites (Chap. 12). This chapter also describes some salient points of difference, in regard to design, between conventional and fiber composite materials. This is indeed of fundamental importance in view of the fact that composite materials are not just any other new material. They represent a total departure from the way we are used to handling conventional monolithic materials, and, consequently, they require unconventional approaches to designing with them.

Throughout this book examples are given from practical applications of composites in various fields. There has been a tremendous increase in applications of composites in sophisticated engineering items. Modern aircraft industry readily comes to mind as an ideal example. Boeing Company, for example, has made widespread use of structural components made of "advanced" composites in 757 and 767 planes. Yet another striking example is that of the Beechcraft Company's Starship 1 aircraft (see frontispiece). This small aircraft (8–10 passengers plus crew) is primarily made of carbon and other high-performance fibers in epoxy matrix. The use of composite materials results in 19% weight reduction compared to an identical aluminum airframe. Besides this weight reduction, the use of composites made a new wing design configuration possible, namely, a variable-geometry forward wing that sweeps forward during takeoff and landing to give stability and sweeps back 30° in level flight to reduce drag. As a bonus, the smooth structure of composite wings helps to maintain laminar air flow. Readers will get an idea of the tremendous advances made in the composites field if they would just remind themselves that until about 1975 these materials were being produced mostly on a laboratory scale. Besides the aerospace industry, chemical, electrical, automobile, and sports industries are the other big users, in one form or another, of composite materials.

This book has grown out of lectures given over a period of more than a decade to audiences comprised of senior year undergraduate and graduate students, as well as practicing engineers from industry. The idea of this book was conceived at Instituto Militar de Engenharia, Rio de Janeiro. I am grateful to my former colleagues there, in particular, J.R.C. Guimarães, W.P. Longo, J.C.M. Suarez, and A.J.P. Haiad, for their stimulating companionship. The book's major gestation period was at the University of Illinois at Urbana–Champaign, where C.A. Wert and J.M. Rigsbee helped me to complete the manuscript. The book is now seeing the light of the day at the New Mexico Institute of Mining and Technology. I would like to thank my colleagues there, in particular, O.T. Inal, P. Lessing, M.A. Meyers, A. Miller, C.J. Popp, and G.R. Purcell, for their cooperation in many ways, tangible and intangible. An immense debt of gratitude is owed to N.J. Grant of MIT, a true gentleman and scholar, for his encouragement, corrections, and suggestions as he read the manuscript. Thanks are also due to R. Signorelli, J. Cornie, and P.K. Rohatgi for reading portions of the manuscript and for their very constructive suggestions. I would be remiss in not mentioning the students who took my courses on composite materials at New Mexico Tech and gave very constructive feedback. An especial mention should be made of C.K. Chang, C.S. Lee, and N. Pehlivanturk for their relentless queries and discussions. Thanks are also due to my wife, Nivedita Chawla, and Elizabeth Fraissinet for their diligent word processing. My son, Nikhilesh Chawla, helped in the index preparation. I would like to express my gratitude to my parents, Manohar L. and Sumitra Chawla, for their ever constant encouragement and inspiration.

Socorro, New Mexico   Krishan Kumar Chawla
June 1987

About the Author

Professor Krishan Kumar Chawla received his B.S. degree from Banaras Hindu University and the M.S. and Ph.D. degrees from the University of Illinois at Urbana–Champaign, all in metallurgical engineering. He has taught and/or done research work at Instituto Militar de Engenharia, Brazil; University of Illinois at Urbana–Champaign; Northwestern University; Université Laval, Canada; and New Mexico Institute of Mining and Technology. He has published extensively in the areas of composite materials and physical and mechanical metallurgy. He is co-author of the text, *Mechanical Metallurgy: Principles & Applications.* Professor Chawla has been a member of the *International Committee on Composite Materials* since its inception in 1975.

Contents

Appendices

Part I

1. Introduction

It is a truism that technological development depends on advances in the field of materials. One does not have to be an expert to realize that a most advanced turbine or aircraft design is of no use if adequate materials to bear the service loads and conditions are not available. Whatever the field may be, the final limitation on advancement depends on materials. Composite materials in this regard represent nothing but a giant step in the ever constant endeavor of optimization in materials.

Strictly speaking, the idea of composite materials is not a new or recent one. Nature is full of examples wherein the idea of composite materials is used. The coconut palm leaf, for example, is nothing but a cantilever using the concept of fiber reinforcement. Wood is a fibrous composite: cellulose fibers in a lignin matrix. The cellulose fibers have high tensile strength but are very flexible (i.e., low stiffness), while the lignin matrix joins the fibers and furnishes the stiffness. Bone is yet another example of a natural composite that supports the weight of various members of the body. It consists of short and soft collagen fibers embedded in a mineral matrix called apatite. A very readable description of the structure–function relationships in the plant and animal kingdoms is available in the book *Mechanical Design in Organisms* [1]. Besides these naturally occurring composites, there are many other engineering materials that are composites in a very general way and have been in use for a long time. The carbon black in rubber, Portland cement or asphalt mixed with sand, and glass fibers in resin are common examples. Thus, we see that the idea of composite materials is not that recent. Nevertheless, one can safely mark the origin of a distinct discipline of composite materials as the beginning of the 1960s. According to a 1973 estimate [2], approximately 80% of all research and development effort in composite materials has been done since 1965. This percentage must have only increased since then. Since the early 1960s, there has been an ever increasing demand for materials ever stiffer and stronger yet lighter in fields as diverse as space, aeronautics, energy, civil construction, etc. The demands made on materials for ever better overall performance are so great and diverse that no one material is able to satisfy them. This naturally led to a resurgence of the ancient concept of combining different materials in an integral–composite material to satisfy the user requirements. Such composite material systems result in a performance unattainable by the individual constituents and they offer the great advantage of a flexible design; that is one can, in principle, tailor-make the material as per specifications of an optimum design. This is a much more powerful statement than might appear at first sight. It implies that, given the most efficient design of, say, an aerospace structure, an automobile, a boat, or an electric motor, we can make a composite material that meets the bill. Schier and Juergens [3] have reviewed the

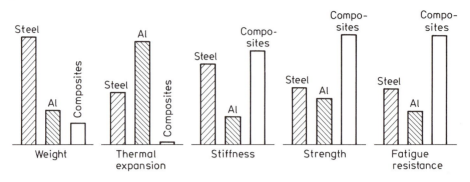

Fig. 1.1. Comparison between conventional monolithic materials and composite materials. (From Ref. 4, used with permission.)

design impact of composites on fighter aircraft. According to these authors, "composites have introduced an extraordinary fluidity to design engineering, in effect forcing the designer-analyst to create a different material for each application as he pursues savings in weight and cost." Yet another conspicuous development has been the integration of the materials science and engineering input with the manufacturing and design inputs at all levels from conception to commissioning of an item and through the inspection during the life time as well as failure analysis. More down-to-earth, however, is the fact that our society has become very energy conscious. This has led to an increasing demand for lightweight yet strong and stiff structures in all walks of life. And composite materials are increasingly providing the answers. Figure 1.1 makes a comparison, admittedly for illustrative purposes, between conventional monolithic materials such aluminum and steel and composite materials [4]. This figure indicates the possibilities of improvements that one can obtain over conventional materials by the use of composite materials. As such it describes vividly the driving force behind the large effort in the field of composite materials.

Glass fiber reinforced resins have been in use since about the first quarter of the twentieth century. Glass fiber reinforced resins are very light and strong materials, although their stiffness (modulus) is not very high, mainly because the glass fiber itself is not very stiff. The third quarter of the twentieth century saw the emergence of the so-called advanced fibers of extremely high modulus, for example, boron, carbon, silicon carbide, and alumina. These fibers have been used for reinforcement of resin, metal, and ceramic matrices. Fiber reinforced composites have been more prominent than other types of composite for the simple reason that most materials are stronger and stiffer in the fibrous form than in any other form. By the same token, it must be recognized that a fibrous form results in reinforcement mainly in fiber direction. Transverse to the fiber direction, there is little or no reinforcement. Of course, one can arrange fibers in two-dimensional or even three-dimensional arrays, but this still does not gainsay the fact that one is not getting the full reinforcement effect in directions other than the fiber axis. Thus, if a less anisotropic behavior is the objective, then perhaps laminate or sandwich composites made of, say, two different metals can be more effective. Here one has a planar reinforcement. There may also be specific nonmechanical objectives for making a fibrous composite. For example, an abrasion or corrosion resistant surface would re-commend the use of a laminate (sandwich) form, while in superconductors the problem of flux-pinning requires the use of extremely fine filaments embedded in a conductive matrix. In what follows, we discuss the various aspects of composites, mostly fiber composites, in greater detail, but first let us agree on an acceptable definition of a composite material.

Practically everything is a composite material in this world. Thus, a common piece of metal is a composite (polycrystal) of many grains (or single crystals). Such a definition would make things quite unwieldy. Therefore, we must agree on an operational definition of composite material for our purposes in this text. We shall call a material that satisfies the following conditions a composite material:

1. It is manufactured (i.e., naturally occurring composites, such as wood, are excluded).
2. It consists of two or more physically and/or chemically distinct, suitably ar-ranged or distributed phases with an interface separating them.
3. It has characteristics that are not depicted by any of the components in isolation.

References

1 S.A. Wainwright, W.D. Biggs, J.D. Currey, and J.M. Gosline, *Mechanical Design in Organisms*, John Wiley & Sons, New York, 1976.
2 H.R. Clauser, *Scientific American*, **229**, 36 (July 1973).
3 J.F. Schier and R.J. Juergens, *Astronautics and Aeronautics*, 44 (Sept. 1983).
4 G.S. Deutsch, *23rd National SAMPE Symposium*, May 1978, p. 34.

2. Fibers

2.1 Introduction

Reinforcements need not necessarily be in the form of long fibers. One can have them in the form of particles, flakes, whiskers, discontinuous fibers, continuous fibers, and sheets. It turns out that the great majority of materials is stronger and stiffer in the fibrous form than in any other form: thus the great attraction of fibrous reinforcements. Specifically, in this category, we are most interested in the so-called advanced fibers which possess very high strength and very high stiffness coupled with a very low density. The reader should realize that many naturally occurring fibers can be and are used in situations involving not very high stresses [1, 2]. The great advantage in this case, of course, is that of low cost. The vegetable kingdom is, in fact, the largest source of fibrous materials. Cellulosic fibers in the form of cotton, flax, jute, hemp, sisal, and ramie, for example, have been used in the textile industry, while wood and straw have been used in the paper industry. Other natural fibers, such as hair, wool, and silk, consist of different forms of protein. Any discussions of such fibers are beyond the scope of this book. The interested reader, however, is directed to a good review article by Meredith [3].

Glass fiber, in its various forms, has been the most common reinforcement for polymer matrices. Kevlar (an aramid) fiber launched by Du Pont in the 1960s is much stiffer and lighter than glass fiber. Other high-performance fibers that combine high strength with high stiffness are boron, silicon carbide, carbon, and alumina. These were all developed in the second part of the twentieth century. In particular, some ceramic fibers were developed in the 1970s and 1980s by a very novel method, namely, the controlled pyrolysis of organic precursors.

The use of fibers as high-performance engineering materials is based on three important characteristics [4]:

1. A small diameter with respect to its grain size or other microstructural unit. This allows a higher fraction of the theoretical strength to be attained than that possible in a bulk form. This is a direct result of the so-called size effect; that is, the smaller the size, the lower the probability of having imperfections in the material. Figure 2.1 shows that the strength of a carbon fiber decreases as its diameter increases [5].

2. A high aspect ratio (length/diameter, l/d) that allows a very large fraction of the applied load to be transferred via the matrix to the stiff and strong fiber (see Chap. 10).

3. A very high degree of flexibility that is really a characteristic of a material having a high modulus and a small diameter. This flexibility permits a variety of techniques to be employed for making composites with these fibers.

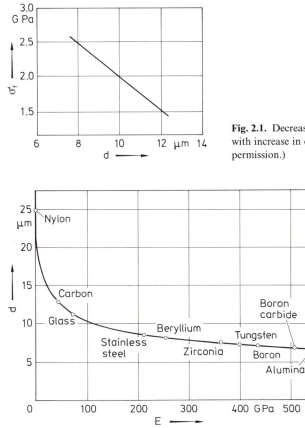

Fig. 2.1. Decrease in strength (σ_f) of a carbon fiber with increase in diameter. (From Ref. 5, used with permission.)

Fig. 2.2. Fiber diameter of different materials with a flexibility equal to that of a 25-µm nylon fiber. (From Ref. 4, used with permission.)

We first consider the concept of flexibility and then go on to describe important fibers in detail. Flexibility of a given material is a function of its elastic modulus E and the moment of inertia of its cross section I. The elastic modulus of a material is quite independent of its form or size. It is generally a constant for a given chemical composition and density. Thus, for a given composition and density, the flexibility of a material is determined by shape, size of the cross section, and its radius of curvature which is a function of its strength. We can use the inverse of the product of bending moment (M) and the radius of curvature (R) as a measure of flexibility. We have

$$\frac{M}{I} = \frac{E}{R}$$

$$MR = EI = \frac{E\pi d^4}{64} \tag{2.1}$$

$$\frac{1}{MR} = \frac{64}{E\pi d^4}$$

where d is the equivalent diameter. Equation (2.1) indicates that $1/MR$, a measure of flexibility, is a very sensitive function of diameter d. Figure 2.2 shows the diameter of various materials in fibrous form which have a flexibility $(1/MR)$ equal to that of a 25-μm diameter nylon fiber (a typical flexible fiber) as a function of the elastic modulus. Note that, given a sufficiently small diameter, it is possible for a metal or ceramic to have the same degree of flexibility as that of a 25-μm diameter nylon. It is another matter that obtaining such a small diameter can be prohibitively expensive.

2.2 Glass Fibers

Glass fiber is a generic name like carbon fiber or steel. A variety of different chemical compositions is commercially available. Common glass fibers are silica based ($\sim$ 50–60% SiO_2) and contain a host of other oxides of calcium, boron, sodium, aluminum, and iron, for example. Table 2.1 gives the compositions of some commonly used glass fibers. The designation E stands for electrical since E glass is a good electrical insulator besides having good strength and a reasonable Young's modulus; C stands for corrosion since C glass has a better resistance to chemical corrosion; S stands for the higher silica content and S glass is able to withstand higher temperatures than other glasses. It should be pointed out that more than 90% of all continuous glass fiber produced is of the E glass type and that, notwithstanding the designation E, electrical uses of E glass fiber are only a small fraction of the total market.

2.2.1 Fabrication

Figure 2.3 shows schematically the conventional fabrication procedure for glass fibers (specifically, the E glass fibers that constitute the workhorse of the resin reinforcement industry) [6–8]. The raw materials are melted in a hopper and the molten glass is fed into the electrically heated platinum bushings or crucibles; each bushing contains 200 holes at its base. The molten glass flows by gravity through these holes forming fine continuous filaments; these are gathered together into a strand and a "size" is applied before it is a wound on a drum. The final fiber diameter is a function of the bushing orifice diameter, viscosity, which is a function of composition and temperature, and the head of glass in the hopper. In many old

Table 2.1. Approximate chemical compositions of some glass fibers (wt.%)

	E Glass	C Glass	S Glass
SiO_2	55.2	65.0	65.0
Al_2O_3	8.0	4.0	25.0
CaO	18.7	14.0	—
MgO	4.6	3.0	10.0
Na_2O	0.3	8.5	0.3
K_2O	0.2	—	—
B_2O_3	7.3	5.0	—

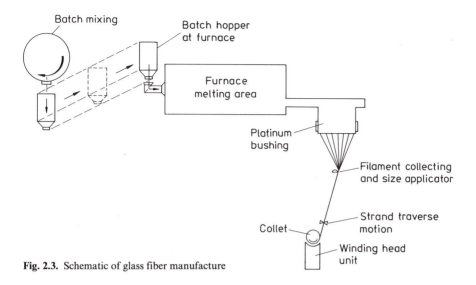

Fig. 2.3. Schematic of glass fiber manufacture

industrial plants the glass fibers are not produced directly from fresh molten glass. Instead, molten glass is first turned into marbles which after inspection are melted in the bushings. Modern plants do produce glass fibers by direct drawing. Figure 2.4 shows some forms in which glass fiber is commercially available.

The conventional methods of making glass or ceramic fibers involve drawing from high-temperature melts of appropriate compositions. This route has many practical difficulties such as the high melting temperatures required, the immiscibility in the liquid state, and the easy crystallization during cooling. Several new techniques were developed in the 1970s for preparing glass and ceramic fibers [9, 10]. One of these processes is called the sol-gel technique. A sol is a colloidal suspension in which the individual particles are so small that they show no sedimentation. A gel is a suspension in which the liquid medium has become viscous enough to behave more or less as a solid. The sol-gel process involves a conversion of fibrous gels, drawn from a solution at near room temperature, into glass or ceramic fibers at several hundred degrees Celsius. The maximum heating temperature in this process is much lower than that in conventional glass fiber manufacture. The sol-gel method using metal alkoxides consists of preparing an appropriate homogeneous solution, changing the solutions to a sol, gelling the sol, and converting the gel to glass by heating. The sol-gel technique is a very powerful technique for making glass and ceramic fibers. The 3M Company produces a silica–alumina fiber, called Nextel, from metal alkoxide solutions (see Sect. 2.6). Figure 2.5 shows an example of drawn silica fibers (cut from a continuous fiber spool) obtained by the sol-gel technique [11].

Glass filaments are easily damaged by the introduction of surface defects. To minimize this and make handling of these fibers easy, a sizing treatment is given. Sizing protects as well as binds the filaments into a strand. For reinforcement purposes, a size based on polyvinyl acetate and containing a resin coupling agent is used; the latter is compatible with polyester, epoxy, and phenolic matrix resins. The coupling agent is used to bond the sized glass fibers and the resin matrix.

Fig. 2.4. Glass fiber is available in a variety of forms: **a** chopped strand, **b** continuous yarn, **c** roving, **d** fabric. (Courtesy of Morrison Molded Fiber Glass Company.)

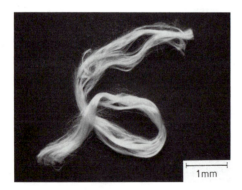

Fig. 2.5. Continuous glass fibers (cut from a spool) obtained by sol-gel technique. (From Ref. 11, used with permission.)

2.2.2 Structure

Glass has an amorphous structure, that is, devoid of any long-range order so characteristic of a crystalline material. Figure 2.6a shows a two-dimensional network of silica glass. Each polyhedron consists of oxygen atoms bonded covalently to silicon. What happens to this structure when Na_2O is thrown in is shown in Fig. 2.6b. Sodium ions are linked ionically with oxygen but they do not join the network

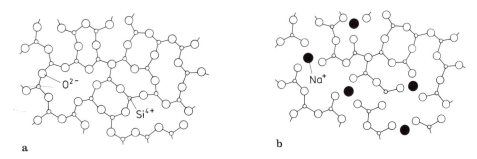

a   b

Fig. 2.6. Amorphous structure of glass: **a** a two-dimensional representation of silica glass network and **b** a modified network that results when Na_2O is added to **a**. Note that Na^+ is ionically linked with O^{2-} but does not join the network directly

directly. Too much Na_2O will impair the tendency for glassy structure formation. The addition of other metal oxide types (Table 2.1) serves to alter the network structure and the bonding and, consequently, the properties. Unlike carbon fibers, glass fibers are quite isotropic; Young's modulus and thermal expansion coefficients are the same along the fiber axis and perpendicular to it. This, of course, is a result of the three-dimensional network structure (Fig. 2.6).

2.2.3 Properties and Applications

Typical mechanical properties of E glass fibers are summarized in Table 2.2. Note that the density is quite low and the strength is quite high; Young's modulus, however, is not very high. Thus, while the strength to weight ratio of glass fibers is quite high, the modulus to weight ratio is only moderate. It is this latter characteristic that led the aerospace industry to other so-called advanced fibers (e.g., boron, carbon, Al_2O_3, and SiC). Glass fibers continue to be used for reinforcement of polyester, epoxy, and phenolic resins. It is quite cheap and is available in a variety of forms (Fig. 2.4). Continuous strand is a group of 204 individual fibers; roving is a group of parallel strands; chopped fibers consist of strand or roving chopped to lengths between 5 and 50 mm. Glass fibers are also available in the form of woven fabrics or nonwoven mats.

    Moisture decreases glass fiber strength. Glass fibers are also susceptible to what has been termed static fatigue; that is, they cannot withstand loads for long periods of time.

Table 2.2. Typical properties of E glass fibers

Density (g cm^{-3})	Tensile strength (MPa)	Young's modulus (GPa)	Coefficient of thermal expansion (K^{-1})
2.55	1750	70	4.7×10^{-6}

Glass fiber reinforced resins are used widely in the building and construction industry. Commonly, these are called glass reinforced plastics or GRP. They are used in the form of a cladding for other structural materials or as an integral part of a structural or nonload bearing wall panel. Window frames, tanks, bathroom units, pipes, and ducts are common examples. Boat hulls, since the mid-1960s, have primarily been made of GRP. Use of GRP in the chemical industry (e.g., as storage tanks, pipelines, and process vessels) is fairly routine now. The rail and road transport industry as well as the aerospace industry are the other big users of GRPs.

2.3 Boron Fibers

Boron is an inherently brittle material. It is commercially made by chemical vapor deposition of boron on a substrate, that is, boron fiber as produced is itself a composite fiber.

In view of the fact that rather high temperatures are required for this deposition process, the choice of substrate material that goes to form the core of the finished boron fiber is limited. Generally, a fine tungsten wire is used for this purpose. A carbon substrate can also been used. The first boron fibers were obtained by Weintraub [12] by means of reduction of a boron halide with hydrogen on a hot wire substrate.

The real impulse in boron fiber fabrication, however, came only in 1959 when Talley [13, 14] used the process of halide reduction to obtain amorphous boron fibers of high strength. Since then, the interest in the use of strong but light boron fibers as a possible structural component in aerospace and other structures has been continuous, although it must be admitted that this interest has periodically waxed and waned in the face of rather stiff competition from other so-called advanced fibers, in particular, carbon fibers.

2.3.1 Fabrication

Boron fibers are obtained by chemical vapor deposition (CVD) on a substrate. There are two processes:

1. *Thermal Decomposition of a Boron Hydride.* This method involves low temperatures, and, thus, carbon coated glass fibers can be used as a substrate. The boron fibers produced by this method, however, are weak because of a lack of adherence between the boron and the core. These fibers are much less dense owing to the trapped gases.
2. *Reduction of Boron Halide.* Hydrogen gas is used to reduce boron trihalide:

$$2\,BX_3 + 3\,H_2 = 2\,B + 6\,HX \tag{2.2}$$

where X denotes a halogen: Cl, Br, or I.

In this process of halide reduction, the temperatures involved are very high, and, thus, one needs a refractory material, for example, a high melting point metal such

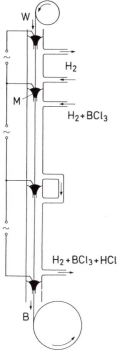

Fig. 2.7. Schematic of boron (B) fiber production by halide decomposition on a tungsten (W) substrate. (From Ref. 15, used with permission.)

as tungsten, as a substrate. It turns out that such metals are also very heavy. This process, however, has won over the thermal reduction process despite the disadvantage of a rather high-density substrate (the density of tungsten is 19.3 g cm^{-3}) mainly because this process gives boron fibers of a very high and uniform quality. There are many firms producing boron fibers commercially using this process. Figure 2.7 shows a schematic of boron filament production while Fig. 2.8 shows a commercial boron filament production facility, each vertical reactor shown in this picture produces continuous boron monofilament.

In the process of BCl_3 reduction, a very fine tungsten wire (10–12 µm diameter) is pulled into a reaction chamber at one end through a mercury seal and out at the other end through another mercury seal. The mercury seals act as electrical contacts for resistance heating of the substrate wire when gases ($BCl_3 + H_2$) pass through the reaction chamber where they react on the incandescent wire substrate. The reactor can be a one- or multistage, vertical or horizontal, reactor. BCl_3 is an expensive chemical and only about 10% of it is converted into boron in this reaction. Thus, an efficient recovery of the unused BCl_3 can result in a considerable lowering of the boron filament cost.

There is a critical temperature for obtaining a boron fiber with optimum properties and structure [15]. The desirable amorphous form of boron occurs below this critical temperature while above this temperature there occur also crystalline forms of boron that are undesirable from a mechanical properties viewpoint, as we shall see in Sect. 2.3.2. With the substrate wire stationary in the reactor, this critical

Fig. 2.8. A boron filament production facility. (Courtesy of AVCO Specialty Materials Co.)

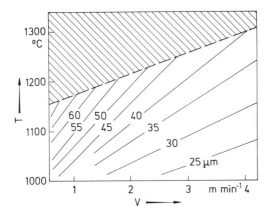

Fig. 2.9. Temperature (T) versus wire speed (V) for a series of boron filament diameters. Filaments formed in the gray region (above the *dashed line*) contain crystalline regions and are undesirable. (From Ref. 15, used with permission.)

temperature is about 1000°C. In a system where the wire is moving, this critical temperature is higher and it increases with the speed of the wire. One generally has a diagram of the type shown in Fig. 2.9, which shows the various combinations of wire temperature and wire drawing speed to produce a certain diameter of boron fiber. Fibers formed in the region above the dashed line are relatively weak because they contain undesirable forms of boron as a result of recrystallization. The explanation for this relationship between critical temperature and wire speed is that boron is deposited in an amorphous state and the more rapidly the wire is drawn out from the reactor, the higher the allowed temperature is. Of course, higher wire drawing speed also results in an increase in production rate and lower costs.

Boron deposition on a carbon monofilament (~ 35 µm diameter) substrate involves precoating the carbon substrate by a layer of pyrolytic graphite. This coating accommodates the growth strains that result during boron deposition [16].

The reactor assembly is slightly different from that for boron on tungsten substrate, since pyrolitic graphite is applied online.

2.3.2 Structure and Morphology

The structure and morphology of boron fibers depend on the conditions of deposition: temperature, composition of gases, gas dynamics, and so on. While theoretically the mechanical properties are limited only by the strength of the atomic bond, in practice, there always are present structural defects and morphological irregularities that lower the mechanical properties. Temperature gradients and trace concentrations of impurity elements inevitably cause process irregularities. Even greater irregularities are caused by fluctuations in electric power, instability in gas flow, or any other operator-induced variables.

Structure

Depending on the conditions of deposition, the elemental boron has been observed in various crystalline polymorphs. The form produced by crystallization from the melt or chemical vapor deposition above 1300°C is β-rhombohedral. At temperatures lower than this, if crystalline boron is produced, the most commonly observed structure is α-rhombohedral.

Boron fibers produced by the CVD method described above have microcrystalline structure that is generally called "amorphous". This designation is based on the characteristic X-ray diffraction pattern produced by the filament in the Debye–Scherrer method, that is, large and diffuse halos with d spacings of 0.44, 0.25, 0.17, 1.4, 1.1, and 0.091 nm, typical of amorphous material [17]. Electron diffraction studies, however, lead one to conclude that this "amorphous" boron is really a microcrystalline phase with grain diameters of the order of 2 nm [16].

Based on X-ray and electron diffraction studies, one can conclude that amorphous boron is really microcrystalline β-rhombohedral. In practice, the presence of microcrystalline phases (crystals or groups of crystals observable in the electron microscope) constitutes an imperfection in the fiber that should be avoided. Larger and more serious imperfections generally result from surpassing the critical temperature of deposition (see Sect. 2.3.1) or the presence of impurities in the gases.

When boron fiber is made by deposition on a tungsten substrate, as is generally the case, then depending on the temperature conditions during deposition, the core may consist of, in addition to tungsten, a series of compounds, such as W_2B, WB, W_2B_5, and WB_4 [18]. A boron fiber cross section (100 μm diameter) is shown in Fig. 2.10a, while Fig. 2.10b shows schematically the various subparts of the cross section. The various tungsten boride phases are formed by diffusion of boron into tungsten. Generally, the fiber core consists only of WB_4 and W_2B_5. On prolonged heating, the core may completely be converted into WB_4. As boron diffuses into the tungsten substrate to form borides, the core expands from its original 12.5 μm (original tungsten wire diameter) to 17.5 μm. The SiC coating shown in Fig. 2.10b is a barrier coating used to prevent any adverse reaction between B and the matrix such as Al at high temperatures. The SiC barrier layer is vapor deposited onto boron using a mixture of hydrogen and methyldichlorosilane.

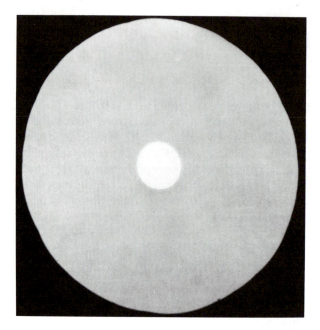

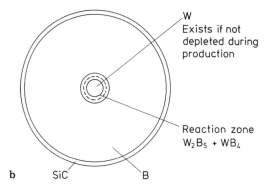

W
Exists if not
depleted during
production

Reaction zone
$W_2B_5 + WB_4$

b   SiC   B

Fig. 2.10. a Cross section of a 100-μm diameter boron fiber. b Schematic of the cross section of a boron fiber with SiC barrier layer

Fig. 2.11. Characteristic corn-cob structure of boron fiber. (From Ref. 15, used with permission.)

Morphology

The boron fiber surface shows a "corn-cob" structure consisting of nodules separated by boundaries (Fig. 2.11). The nodule size varies during the course of fabircation. In a very general way, the nodules start as individual nuclei on the substrate and then grow outward in a conical form until a filament diameter of

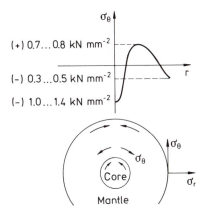

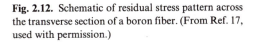

Fig. 2.12. Schematic of residual stress pattern across the transverse section of a boron fiber. (From Ref. 17, used with permission.)

80–90 μm is reached, above which the nodules seem to decrease in size. Occasionally, new cones may nucleate in the material, but they always originate at an interface with a foreign particle or inclusion.

2.3.3 Residual Stresses

Boron fibers have inherent residual stresses that have their origin in the process of chemical vapor deposition. Growth stresses in the nodules of boron, stresses induced by the diffusion of boron into the tungsten core, and stresses generated by the difference in the coefficient of expansion of deposited boron and tungsten boride core, all contribute to the residual stresses and thus can have a considerable influence on the fiber mechanical properties. A schematic of the residual stress pattern across the transverse section of a boron fiber is shown in Fig. 2.12 [17]. The compressive stresses on the fiber surface are due to the quenching action involved in pulling the fiber out from the chamber [17]. Morphologically, the most conspicuous aspect of these internal stresses would appear to be the frequently observed radial crack, from within the core to just inside the outer surface, in the transverse section of these fibers. Some workers, however, doubt the preexistence of this radial crack [16]. They think that the crack appears during the process of boron fiber fracture.

2.3.4 Fracture Characteristics

It is well known that brittle materials show a distribution of strengths rather than a single value. Imperfections in these materials lead to stress concentrations much higher than the applied stress levels. Because the brittle material is not capable of deforming plastically in response to these stress concentrations, fracture ensues at one or more such sites. Boron fiber is indeed a very brittle material and cracks originate at preexisting defects located at the boron–core interface or at the surface. Figure 2.13 shows the characteristic brittle fracture of a boron fiber and the radial crack. The surface defects are due to the nodular surface which results from the growth of boron cones. In particular, when a nodule coarsens due to an exaggerated

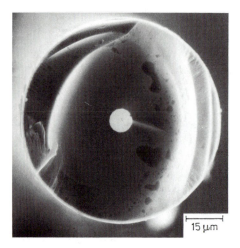

15 μm

Fig. 2.13. Fracture surface of a boron fiber showing a characteristically brittle fracture and a radial crack

growth around a contaminating particle, a crack can result from this large nodule and weaken the fiber.

As mentioned in the beginning, boron fiber in itself is a composite fiber. It is a consequence of the discontinuity between the properties of boron and tungsten borides. This discontinuity cannot be eliminated totally but can be minimized by a proper core formation and proper bonding between the core and the boron deposit.

2.3.5 Properties and Applications of Boron Fibers [16–22]

Due to the composite nature of the boron fiber, complex internal stresses and defects such as voids and structural discontinuities result from the presence of a core and the deposition process. Thus, one would not expect boron fiber strength to equal the intrinsic strength of boron. The average tensile strength of boron fiber is 3–4 GPa, while its Young's modulus is between 380 and 400 GPa.

An idea of the intrinsic strength of boron is obtained in a flexure test [21]. It would be expected that in flexure, assuming the core and interface to be near the neutral axis, critical tensile stresses would not develop at the core or interface. Flexure tests on boron fibers lightly etched to remove any surface defects gave a strength of 14 GPa. Without etching the strength was half this value.

There has been some effort at NASA Lewis Research Center to improve the tensile strength and toughness (or fracture energy) of boron fibers by making them larger in diameter [22]. We shall discuss the requirements for metal matrix composites (MMC) in terms of the fiber diameter and other parameters in Chapter 6. Here, we restrict ourselves to boron fibers. Commercially produced 142-μm diameter boron fiber shows tensile strengths less than 3.8 GPa. The tensile strength and fracture energy values of the as-received and some limited-production run larger-diameter fibers showed improvement after chemical polishing as shown in Table 2.3. Fibers showing strengths above 4 GPa had their fracture controlled by a tungsten–boride core, while fibers with strengths of 4 GPa were controlled by

Table 2.3. Strength properties of improved large-diameter boron fibers

Diameter	Treatment	Strength		Relative fracture energy
		Average[a] (GPa)	COV[b] (%)	
(μm)				
142	As-produced	3.8	10	1.0
406	As-produced	2.1	14	0.3
382	Chemical polish	4.6	4	1.4
382	Heat treatment plus polish	5.7	4	2.2

[a] Gauge length = 25 mm.
[b] Coefficient of variation = standard deviation/average value.
Source: Reprinted with permission from *Journal of Metals,* **37**, No. 6, 1985, a publication of The Metallurgical Society, Warrendale, Pennsylvania.

fiber surface flaws. The high-temperature treatment, listed in Table 2.3, improved the fiber properties by putting a permanent axial contraction strain in the sheath.

Boron has a density of 2.34 g cm^{-3} (about 15% less than that of aluminum). Boron fiber with the tungsten core has a density of 2.6 g cm^{-3} for a fiber of 100 μm diameter. Its melting point is 2040°C and it has a thermal expansion coefficient of 8.3×10^{-6} °C^{-1} up to 315°C.

Boron fiber composites are in use in a number of U.S. military aircraft, notably the F-14 and F-15, and in the U.S. Space Shuttle. Increasingly, boron fibers are being used for stiffening golf shafts, tennis rackets, and bicycle frames. One big obstacle to the widespread use of boron fiber is its high cost compared to that of other fibers. A major portion of this high price is the cost of the tungsten substrate.

2.4 Carbon Fibers

Carbon is a very light element with a density equal to 2.268 g cm^{-3}. Carbon can exist in a variety of crystalline forms. Our interest in the present case is in the so-called graphitic structure wherein the carbon atoms are arranged in the form of hexagonal layers. The other well-known form of carbon is the covalent diamond structure wherein the carbon atoms are arranged in a three-dimensional configuration with little structural flexibility. Carbon in the graphitic form is highly anisotropic with a Young's modulus in the layer plane being equal to about 1000 GPa while that along the c axis is equal to about 35 GPa. The graphite structure, Fig. 2.14a, has a very dense packing in the layer planes. The lattice structure is shown more clearly with only lattice planes in Fig. 2.14b. As we know, the bond strength determines the modulus of a material. Thus, the high-strength bond between carbon atoms in the layer plane results in an extremely high modulus while the weak van der Waals type bond between the neighboring layers results in a lower modulus in that direction. Consequently, one would like to have in a carbon fiber a very high degree of preferred orientation of hexagonal planes along the fiber axis.

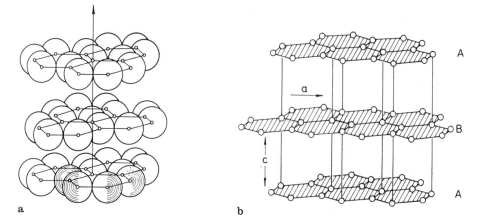

Fig. 2.14. a The densely packed graphitic layer structure. **b** The hexagonal lattice structure of graphite

Carbon fibers of extremely high modulus can be made by carbonization of organic precursor fibers followed by graphitization at high temperatures. The organic precursor fiber, that is, the raw material for carbon fiber, is generally a special textile polymeric fiber that can be carbonized without melting. The precursor fiber, like any polymeric fiber, consists of long-chain molecules (0.1–1 μm when fully stretched) arranged in a random manner. Such polymeric fibers generally have poor mechanical properties and typically show rather large deformations at low stresses mainly because the polymeric chains are not ordered. The commonly used precursor fibers are polyacrylonitrile (PAN) and rayon. Other precursors are obtained from pitches, polyvinylalcohol, polyimides, and phenolics.

The name carbon fiber is a generic one representing a family of fibers [23]. As pointed out above, unlike the rigid diamond structure, graphitic carbon has a lamellar structure. Thus, depending on the size of the lamellar packets, their stacking heights, and the resulting crystalline orientations, one can obtain a range of properties. Most of the carbon fiber fabrication processes involve the following essential steps:

1. A *stabilization* treatment that prevents the fiber from melting in the subsequent high-temperature treatments.
2. A thermal treatment called *carbonization* that removes a great majority of noncarbon elements.
3. An optional thermal treatment called *graphitization* that improves the properties of carbon fiber obtained in step 2.

It should be clear to the reader by now that to have a high modulus fiber, one needs to improve the orientation of graphitic crystals or lamellas. This is achieved by various kinds of thermal and stretching treatments involving rather rigorous controls. If a constant stress were applied, for example, it would result in excessive fiber elongation and the accompanying reduction in area could lead to fiber fracture.

2.4.1 Preparation

Shindo in Japan was the first one to prepare high-modulus carbon fiber starting from PAN in 1961 [24]. He obtained a Young's modulus of about 170 GPa. In 1963, British researchers at Rolls Royce discovered that a high elastic modulus of carbon fiber was obtained by means of stretching. They obtained, starting from PAN, a carbon fiber with modulus of about 600 GPa. Since then, developments have occurred in rapid strides in the technology of carbon fibers. The minute details of the conversion processes from precursor fiber to a high-modulus carbon fiber continue to be proprietary secrets. All the methods, however, exploit the phenomenon of thermal decomposition of an organic fiber under well-controlled conditions of rate and time of heating, environment, and so on. Also, in all processes the precursor is stretched at some stage of pyrolysis to obtain the high degree of alignment of graphitic basal planes.

PAN Precursor

The polyacrylonitrile fibers are stabilized in air (a few hours at 250°C) to prevent melting during the subsequent higher-temperature treatment. The fibers are prevented from contracting during this oxidation treatment. The black fibers obtained after this treatment are heated slowly in an inert atmosphere to 1000–1500°C. Slow heating allows the high degree of order present in the fiber to be maintained. The rate of temperature increase should be low so as not to destroy the molecular order present in the fibers. The final heat treatment consists of holding the fibers for very short duration at temperatures up to 3000°C. This improves the fiber texture and thus increases the elastic modulus of the fiber. Figure 2.15 shows, schematically, this PAN-based carbon fiber production process [25]. Typically, the carbon fiber yield is about 50%.

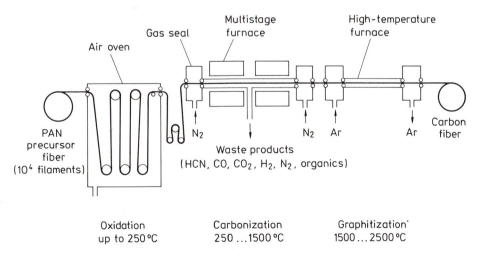

Fig. 2.15. Schematic of PAN-based carbon fiber production. (Reprinted with permission from [Metals Forum, 6, A.A. Baker, 81], Copyright [1983], Pergamon Press, Ltd.)

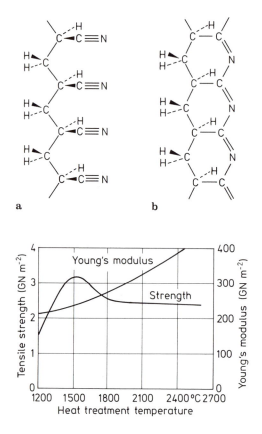

Fig. 2.16. a Flexible polyacrylonitrile molecule. **b** Rigid ladder (or oriented cyclic) molecule

Fig. 2.17. Strength and elastic modulus of carbon fiber as a function of final heat treatment temperature. (After Ref. 26, used with permission.)

The initial stretching treatment of PAN improves the axial alignment of the polymer molecules. During this oxidation treatment the fibers are maintained under tension to keep the alignment of PAN while it transforms into rigid ladder polymer (Fig. 2.16). In the absence of this tensile stress in this step, there will occur a relaxation and the ladder polymer structure will become disoriented with respect to the fiber axis. After the stabilizing treatment, the resulting ladder type structure (also called oriented cyclic structure) has a high glass transition temperature so that there is no need to stretch the fiber during the next stage, namely, carbonization. There still are present considerable quantities of nitrogen and hydrogen. These are eliminated as gaseous waste products during carbonization, that is, heating to 1000–1500°C (Fig. 2.15). The carbon atoms remaining after this treatment are mainly in the form of a network of extended hexagonal ribbons which has been called *turbostratic* graphite structure in the literature. Although these strips tend to align parallel to the fiber axis, the degree of order of one ribbon with respect to another is relatively low. This can be improved further by heat treatment at still higher temperatures (up to 3000°C). This is called the graphitization treatment (Fig. 2.15). The mechanical properties of the resultant carbon fiber may vary over a large range depending mainly on the temperature of the final heat treatment (Fig. 2.17) [26]. Hot stretching above 2000°C results in plastic deformation of fibers leading to an improvement in properties.

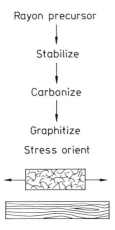

Rayon precursor

Stabilize

Carbonize

Graphitize
Stress orient

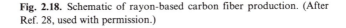

Fig. 2.18. Schematic of rayon-based carbon fiber production. (After Ref. 28, used with permission.)

Cellulosic Precursors

Cellulose is a natural polymer and is frequently found in a fibrous form. In fact, cotton fiber, which is cellulosic, was one of the first ones to be carbonized. It has the desirable property of decomposing before melting. It is inappropriate, however, for high-modulus carbon fiber manufacture because it has a rather low degree of orientation along the fiber axis, although it is highly crystalline. It is also not available as a tow of continuous filaments and is quite expensive. These difficulties have been overcome in the case of rayon fiber, which is made from wood pulp, a cheap source. The cellulose is extracted from wood pulp and continuous filament tows are produced by wet spinning.

Rayon is a thermosetting polymer. The process used for the conversion of rayon into carbon fiber involves the same three stages: stabilization in a reactive atmosphere (air or oxygen, $<400°C$), carbonization ($<1500°C$), and graphitization ($>2500°C$). Various reactions occur during the first stage, causing extensive decomposition and evolution of H_2O, CO, CO_2, and tar. The stabilization is carried out in a reactive atmosphere to inhibit tar formation and improve yield [27]. Chain fragmentation or depolymerization occurs in this stage. Because of this depolymerization, stabilizing under tension, as done in the case of PAN precursor, does not work in this case [27]. The carbonization treatment involves heating to about 1000°C in nitrogen. Graphitization is carried out at 2800°C but under stress. This orienting stress at high temperature results in plastic deformation via multiple slip system operation and diffusion. Figure 2.18 shows the process schematically. The carbon fiber yield from rayon is between 15 and 30% by weight compared to a yield of about 50% in the case of PAN precursors.

Pitch-Based Carbon Fibers

There are various sources of pitch but the three commonly used sources are polyvinyl chloride (PVC), petroleum asphalt, and coal tar. Pitch-based carbon fibers have become attractive because of the cheap raw material and high yield of carbon fibers.

The same sequence of oxidation, carbonization, and graphitization is required for making carbon fibers out of pitch precursors. Orientation in this case is obtained by spinning. An isotropic but aromatic pitch is subjected to melt spinning at very high strain rates and quenched to give a highly oriented fiber. This thermoplastic fiber is then oxidized to form a crosslinked structure that makes the fiber nonmelting. This is followed by carbonization and graphitization.

Commercial pitches are mixtures of various organic compounds with an average molecular weight between 400 and 600. Prolonged heating above 350°C results in the formation of a highly oriented, optically anisotropic liquid crystalline phase (mesophase). When observed under polarized light, anisotropic mesophase dispersed in an isotropic pitch appears as microspheres floating in pitch. The liquid crystalline mesophase pitch can be melt spun into a precursor for carbon fiber. The melt spinning process involves shear and elongation in the fiber axis direction and thus a high degree of preferred orientation is achieved. This orientation can be further developed during conversion to carbon fiber. The pitch molecules (aromatics of low molecular weight) are stripped of hydrogen and the aromatic molecules coalesce to form larger bidimensional molecules. Very high values of Young's modulus can be obtained. It should be appreciated that one must have the pitch in a state amenable to spinning in order to produce the precursor fiber. This precursor fiber is made infusible to allow carbonization to occur without melting. Thus, the pitches obtained from petroleum asphalt and coal tar need pretreatments. This pretreatment can be avoided in the case of PVC by means of a carefully controlled thermal degradation of PVC. The molecular weight controls the viscosity of the melt polymer and the melting range. Thus, it also controls the temperature and the spinning speed. Because the pitches are polydispersoid systems, their molecular weights can be adjusted by solvent extraction or distillation. Figure 2.19 shows the process of pitch-based carbon fiber manufacture starting from an isotropic pitch and a mesophase pitch [28].

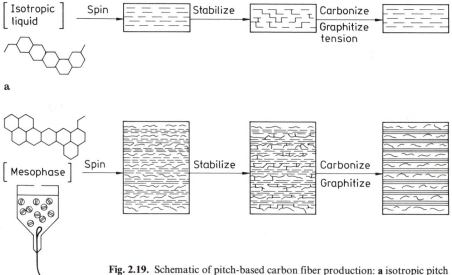

Fig. 2.19. Schematic of pitch-based carbon fiber production: **a** isotropic pitch process, **b** mesophase pitch process. (After Ref. 28, used with permission.)

2.4.2 Structural Changes Occurring During Processing

The thermal treatments for all precursor fibers serve to remove noncarbon elements in the form of gases. For this, the precursor fibers are stabilized (they become black) to ensure that they decompose before melting. Carbon fibers obtained after carbonization contain many "grown in" defects because the thermal energy supplied at these low temperatures is not enough to break the already formed carbon–carbon bonds. That is why these carbon fibers are very stable up to 2500–3000°C when they change to graphite. The decomposition of the precursor fiber invariably results in a weight loss and a decrease in fiber diameter. The weight loss can be considerable — from 40 to 90% depending on the precursor and treatment [29]. The external morphology of the fiber, however, is generally maintained. Thus, precursor fibers with transverse sections in the form of a kidney bean, dog bone, or circle maintain this form after conversion to carbon fiber. Figure 2.20 shows a scanning electron micrograph of a PAN-based carbon fiber.

At the microscopic level, carbon fibers possess a rather heterogeneous microstructure. Not surprisingly, many workers [28, 30–35] have attempted to characterize the structure of carbon fibers and there are in the literature a number of models. There exists a better understanding of the structure of PAN-based carbon fibers. Essentially, a carbon fiber consists of many graphitic lamellar ribbons oriented roughly parallel to the fiber axis with a complex interlinking of layer planes both longitudinally and laterally. Based on high-resolution lattice fringe images of longitudinal and transverse sections in TEM, a schematic two-dimensional representation is given in Fig. 2.21 [34], while a three-dimensional model is shown in Fig. 2.22 [33]. The structure is typically defined in terms of crystallite dimensions, L_a and L_c in directions a and c, respectively, as shown in Fig. 2.23. The degree of alignment as well as the parameters L_a and L_c vary with the graphitization temperature. Both L_a and L_c increase with increasing heat treatment temperature. There always are present flaws of various kinds which may arise from impurities in the precursor or may simply be the misoriented layer planes. A mechanism of tensile failure of carbon fiber based on the presence of misoriented crystallites is shown in Fig. 2.23 [36]. Figure 2.23a shows a misoriented crystallite linking two crystallites parallel to the fiber axis. Under the action of applied stress, basal plane rupture occurs in the misoriented crystallite in the L_c direction, followed by crack development along L_a and L_c (Fig. 2.23b). Continued stressing causes complete failure of the misoriented crystallite (Fig. 2.23c). If the crack size is greater than the critical size in the L_a and L_c directions, catastrophic failure results.

Fig. 2.20. Scanning electron micrograph of PAN-based carbon fiber (fiber diameter is 8 μm)

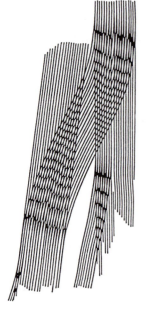
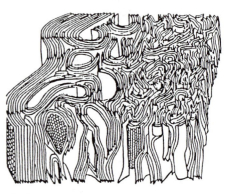

Fig. 2.21   **Fig. 2.22**

Fig. 2.21. Two-dimensional representation of PAN-based carbon fiber. (Reprinted with permission from [Carbon, 17, S.C. Bennett and D.J. Johnson, 25], Copyright [1979], Pergamon Press, Ltd.)

Fig. 2.22. Three-dimensional representation of PAN-based carbon fiber. (From Ref. 33, used with permission.)

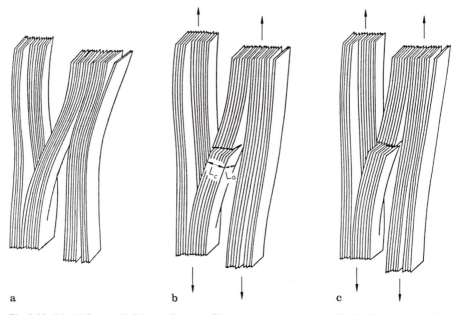

a   b   c

Fig. 2.23. Model for tensile failure of carbon fiber: **a** a misoriented crystallite linking two crystallites parallel to the fiber axis, **b** basal plane rupture under the action of applied stress, **c** complete failure of the misoriented crystallite. (From Ref. 36, used with permission.)

2.4.3 Properties and Applications

The density of the carbon fiber varies with the precursor and the thermal treatment given. It varies in the range of 1.6–2.2 g cm^{-3}. Note that the density of the carbon fiber is more than that of the precursor fiber; the density of the precursor being generally between 1.14 and 1.19 g cm^{-3} [36].

As mentioned above, the degree of order, and consequently the modulus in the fiber axis direction, increases with increasing graphitization temperature. Fourdeux et al. [37] measured the preferred orientation of various carbon fibers and plotted it against an orientation parameter q (Fig. 2.24). The parameter q has a value of -1 for perfect orientation and zero for the isotropic case. In Fig. 2.24 we have plotted the absolute value of q. Note also that the modulus has been corrected for porosity. The theoretical curve fits the experimental data very well.

Even among the PAN carbon fibers, we can have a series of carbon fibers: for example, high tensile strength but medium Young's modulus (HT) fiber (200–300 GPa); high Young's modulus (HM) fiber (400 GPa); extra- or superhigh tensile strength (SHT); and superhigh modulus type (SHM) carbon fibers. The mesophase pitch-based carbon fibers show rather high modulus but low strength levels (~ 2 GPa). Not unexpectedly, the HT type carbon fibers show a much higher strain to failure value than the HM type. The former are more widely used. The mesophase pitch-based carbon fibers are used for reinforcement, while the isotropic-based fibers are more frequently used as insulation and fillers. Table 2.4 compares the properties of some commonly obtainable carbon fibers and graphite monocrystal [38]. For high-temperature applications involving carbon fibers, it is important to take into account the variation of inherent oxidation resistance of carbon fibers with modulus. Figure 2.25 shows that the oxidation resistance of carbon fiber increases with the modulus value [39]. The modulus, as we know, increases with the final heat treatment temperature during processing.

We note from Table 2.4 that the carbon fibers produced from various precursor materials are fairly good electrical conductors. Although this had led to some work toward a potential use of carbon fibers as current carriers for electrical power transmission [40], it has also caused extreme concern in many quarters. The reason for this concern is that if the extremely fine carbon fibers accidentally become

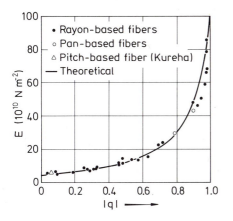

Fig. 2.24. Modulus variation for various carbon fibers with the degree of preferred orientation. The value of orientation parameter, q, is 1 for perfect orientation and zero for the isotropic case. (From Ref. 37, used with permission.)

Table 2.4. Comparison of properties of different carbon fibers

Precursor	Density (g cm^{-3})	Young's modulus (GPa)	Electrical resistivity (10^{-4} Ω cm)
Rayon[a]	1.66	390	10
Polyacrylonitrile[b] (PAN)	1.74	230	18
Pitch (Kureha)			
LT[c]	1.6	41	100
HT[d]	1.6	41	50
Mesophase pitch[e]			
LT	2.1	340	9
HT	2.2	690	1.8
Single-crystal[f]			
graphite	2.25	1000	0.40

[a] Union Carbide, Thornel 50.
[b] Union Carbide, Thornel 300.
[c] LT, low-temperature heat-treated.
[d] HT, high-temperature heat-treated.
[e] Union Carbide type P fibers.
[f] Modulus and resistivity are in-plane values.
Source: Adapted with permission from Ref. 38.

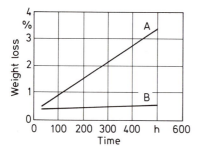

Fig. 2.25. Oxidation resistance, measured as weight loss in air at 350°C, of carbon fibers having different moduli: (A) Celion 3000 (240 GPa) and (B) Celion G-50 (345 GPa). (After J.P. Riggs, Encyclopedia of Polymer Science & Engineering, 2nd. ed., vol. 2, copyright © 1985, John Wiley & Sons, New York. Reprinted by permission of John Wiley & Sons, Inc.)

airborne (during manufacture or service) they can settle on electrical equipment and cause short circuiting.

Anisotropic as the carbon fibers are, they have two principal coefficients of thermal expansion, namely, transverse or perpendicular to the fiber axis, α_t, and parallel to the fiber axis α_l. Typical values are

$$\alpha_t \simeq \quad 5.5 \text{ to} \quad 8.4 \times 10^{-6} \text{ K}^{-1}$$

$$\alpha_l \simeq -0.5 \text{ to} -1.3 \times 10^{-6} \text{ K}^{-1}$$

Carbon fibers have found a variety of applications in the aerospace and sporting goods industries. Cargo bay doors and booster rocket casings in the U.S. shuttle are made of carbon fiber reinforced epoxy composites. Modern commerical aircrafts also use carbon fiber reinforced composites. Among other areas of application of carbon fibers, one can cite various machinery items such as turbine, compressor, and windmill blades and flywheels; in the field of medicine the applications include both equipment as well as implant materials (e.g., ligament replacement in knees and hip joint replacement). We discuss these in more detail in Chap. 8.

2.5 Organic Fibers

In view of the fact that the covalent carbon–carbon bond is a very strong one, we should expect linear chain polymers such as polyethylene to be potentially very strong and stiff. What one needs for realizing this potential is full extension of molecular chains. The orientation of these polymer chains with respect to the fiber axis and the manner in which they fit together (i.e., order or crystallinity) are controlled by their chemical nature and the processing route. There are two ways of achieving molecular orientation, one without high molecular extension (Fig. 2.26a) and the other with high molecular extension (Fig. 2.26b). During the 1970s and 1980s considerable effort has gone toward realizing this potential in the simple linear polymer polyethylene and impressive results have been obtained on a laboratory scale. Allied Corporation announced in the mid-1980s an extended chain ultrahigh molecular weight (UHMW) polyethylene (PE) fiber, trade name Spectra 900, with impressive properties. The other highly successful organic fiber is aramid fiber, commercialized by Du Pont under the trade name Kevlar. In what follows we describe very briefly the impressive results obtained with polyethylene fibers and then describe the aramid fibers in more detail.

2.5.1 Oriented Polyethylene Fibers

Barham and Keller have reviewed the work on obtaining highly oriented polyethylene fibers [41]. In the mid-1970s reports of producing strong and stiff polyethylene fibers [42] started to appear. Most of this work involved drawing of melt crystallized polyethylene to very high draw ratios [30]. Polyethylene of molecular weights of 10^4–10^5 were used and moduli of up to 70 GPa were obtained. Tensile drawing, die drawing, or hydrostatic extrusion were used to obtain the high plastic strains required for obtaining a high modulus. It turns out that modulus is dependent on the draw ratio, independent of how the draw ratio is obtained [42]. In all these drawing processes, the polymer chains become merely oriented without undergoing molecular extension (Fig. 2.26a).

Later developments have involved altogether different processing routes, solution and gel spinning of very high molecular weight polyethylenes ($> 10^6$) which

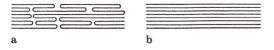

a   b

Fig. 2.26. Two ways of achieving molecular orientation: **a** without high molecular extension and **b** with high molecular extension. (From Ref. 41, used with permission.)

resulted in moduli as high as 200 GPa. In all these methods, molecular orientation is achieved together with chain extension (Fig. 2.26b). In one method, lamellar single crystals of polyethylene are obtained from the solution during crystallization, aligning the molecular chains in the direction of flow. In another method, a crystallized "shish kebab" structure is obtained from the polymer solution. The chains are quite extended in this structure. A shish kebab structure consists of a continuous array of fibrous crystals, the shish kebabs, in which the molecular chains are highly extended. The third method of crystallization, and perhaps technologically the most important, leads to gels. Gels are nothing but swollen networks in which crystalline regions form the junctions. Essentially, an appropriate polymer solution is converted into gel which can be processed by a variety of methods to give the fiber. High molecular weight of the polymer and high concentration of the solution for a given molecular weight promote gel-forming crystallization. The alignment and extension of chains is obtained by the drawing of gel fiber. One problem with this gel route is the rather low spinning rates of 1.5 m min^{-1}. At higher rates, the properties obtained are not very good [43, 44]. Allied Corporation launched in the mid-1980s an UHMW-PE fiber, called Spectra 900, obtained by the gel processing route. Spectra 900 fiber is very light with a density of 0.97 g cm^{-3}. Its strength and modulus are slightly lower than those of aramid fibers but on a per unit weight basis, Spectra 900 has values about 30–40% higher than those of Kevlar. It should be pointed out that both those fibers, as is true of most organic fibers, must be limited to low-temperature applications. Spectra 900, for example, melts at 150°C. This solution spinning approach to producing high-modulus and high-strength fibers has been successfully applied in producing the aramid fibers. We describe these in the next section.

2.5.2 Aramid Fibers

Aramid fiber is a generic name for a class of synthetic organic fibers called aromatic polyamide fibers. The U.S. Federal Trade Commission gives a good definition of an aramid fiber as "a manufactured fiber in which the fiber forming substance is a long chain synthetic polyamide in which at least 85% of the amide linkages are attached directly to two aromatic rings." Researchers at the Monsanto and Du Pont companies were independently able to produce high-modulus aromatic fibers. Only Du Pont, however, has produced them commercially under the trade name Kevlar since 1971.

Nylon is a generic name for any long-chain polyamide. Aramid fibers like Nomex or Kevlar, however, are ring compounds based on the structure of benzene as opposed to linear compounds used to make nylon. The basic chemical structure of aramid fibers consists of oriented para-substituted aromatic units, which makes them rigid rodlike polymers. The rigid rodlike structure results in a high glass transition temperature and poor solubility, which makes fabrication of these polymers, by conventional drawing techniques, difficult. Instead, they are melt spun from liquid crystalline polymer solutions as described below.

Fabrication

Although the specific details of the manufacturing of aramid fibers remain proprietary secrets, it is believed that the processing route involves solution–polycondensation of diamines and diacid halides at low temperatures. Hodd and Turley [45], Morgan [46], and Magat [47] have given simplified accounts of the theory involved in the fabrication of aramid fibers. The most important point is that the starting spinnable solutions that give high-strength and high-modulus fibers have liquid crystalline order. Figure 2.27 shows schematically various states of a polymer in solution. Figure 2.27a shows two-dimensional, linear, flexible chain polymers in solution. These are called random coils as the figure suggests. If the polymer chains can be made of rigid units, that is, rodlike, we can represent them as a random array of rods (Fig. 2.27b). Any associated solvent may contribute to the rigidity and to the volume occupied by each polymer molecule. With increasing concentration of rodlike molecules, one can dissolve more polymer by forming regions of partial order, that is, regions in which the chains form a parallel array. This partially ordered state is called a liquid crystalline state (Fig. 2.27c). When the rodlike chains become approximately arranged parallel to their long axes but their centers remain unorganized or randomly distributed, we have what is called a *nematic* liquid crystal (Fig. 2.27d). It is this kind of order that is found in the extended chain polyamides.

Liquid crystal solutions, because of the presence of the ordered domains, are optically anisotropic, that is, birefringent. Figure 2.28 shows the anisotropic Kevlar

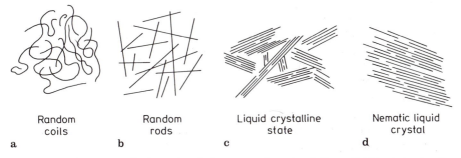

| Random coils | Random rods | Liquid crystalline state | Nematic liquid crystal |
| a | b | c | d |

Fig. 2.27. Various states of polymer in solution: **a** two-dimensional, linear, flexible chains (random coils), **b** random array of rods, **c** partially ordered liquid crystalline state, and **d** nematic liquid crystal (randomly distributed parallel rods)

Fig. 2.28. Anisotropic Kevlar aramid and sulfuric acid solution at rest between crossed polarizers. (Courtesy of Du Pont Co.)

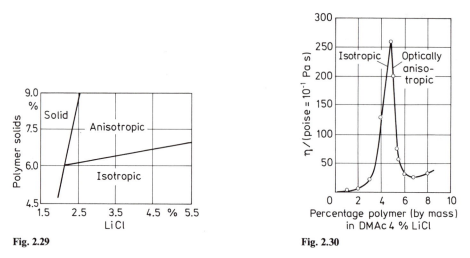

Fig. 2.29

Fig. 2.30

Fig. 2.29. Phase diagram of poly-*p*-benzamide in tetramethylurea—LiCl solutions. Note that the anisotropic state is obtained under certain conditions. (From Ref. 47, used with permission.)

Fig. 2.30. Viscosity versus polymer concentration in solution. A sharp drop in viscosity occurs where
the solution starts becoming anisotropic liquid crystal. (From Ref. 47, used with permission.)

aramid and sulfuric acid solution at rest between crossed polarizers. The parallel
arrays of polymer chains in the liquid crystalline state become even more ordered
when these solutions are subjected to shear as, for example, in extruding through a
spinneret hole. It is this inherent property of liquid crystal solutions which is
exploited in the manufacture of aramid fibers. The characteristic fibrillar structure
of aramid fibers is due to the alignment of polymer crystallites along the fiber axes.

Para-oriented aromatic polyamides form liquid crystal solutions under certain
conditions of concentration, temperature, solvent, and molecular weight. Figure
2.29 shows a phase diagram of the system poly-*p*-benzamide in tetramethylurea—
LiCl solutions [47]. Only under certain conditions do we get the desirable anisotropic state. There also occurs an anomalous relationship between viscosity and
polymer concentration in liquid crystal solutions. Initially, there occurs an increase in viscosity as the concentration of polymer in solution increases, as it would
in any ordinary polymer solution. At a critical point where it starts assuming an
anisotropic liquid crystalline shape, there occurs a sharp drop in the viscosity; see
Fig. 2.30 [47]. The liquid crystalline regions act like dispersed particles and do not
contribute to solution viscosity. With increasing polymer concentration, the
amount of liquid crystalline phase increases up to a point when the viscosity tends to
rise again. There are other requirements for forming a liquid crystalline solution
from aromatic polyamides. The molecular weight must be above some minimum
value and the solubility must exceed the critical concentration required for liquid
crystallinity. Thus, starting from liquid crystalline spinning solutions containing
highly ordered arrays of extended polymer chains, we can spin fibers directly into
an extremely oriented, chain-extended form. These as-spun fibers are quite strong
and, since the chains are highly extended and oriented, one does not need to use
conventional drawing techniques.

Researchers at Du Pont [48] discovered a spinning solvent for poly-*p*-benzamide (PBA) and were able to dry spin quite strong fibers from tetramethylurea–LiCl solutions. This was the real breakthrough. The modulus of these as-spun organic fibers was greater than that of glass fibers. *p*-Oriented rigid diamines and dibasic acids give polyamides that yield, under appropriate conditions of solvent, concentration, and polymer molecular weight, the desired nematic liquid crystal structure. One would like to have, for any solution spinning process, a high molecular weight to obtain improved mechanical properties, a low viscosity to ease processing conditions, and a high polymer concentration to achieve a high yield. For *para*-aramid, poly *p*-phenyleneterephthalamide (PPD-T), trade name Kevlar, the nematic liquid crystalline state is obtained in 100% sulfuric acid at a polymer concentration of about 20% (49). The polymer solution is often referred to as the dope. The various spinning processes available are classified as dry, wet, and dry jet–wet spinning processes. For aramid fibers, the dry jet–wet spinning method is employed [50]. The process is illustrated in Fig. 2.31. It is believed that solution–polycondensation of diamines and diacid halides at low temperatures (near 0°C) gives the aramid-forming polyamides. Low temperatures inhibit by-product generation and promote linear polyamide formation. The resulting polymer is pulverized, washed, and dried. This is mixed with a strong acid (e.g., concen-

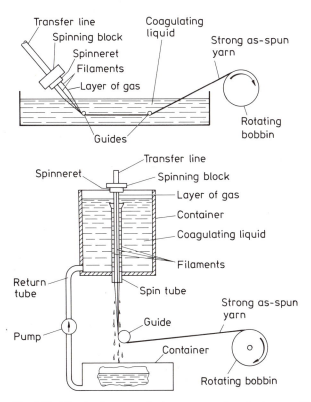

Fig. 2.31. The dry jet–wet spinning process of producing aramid fibers. (After Ref. 50, used with permission.)

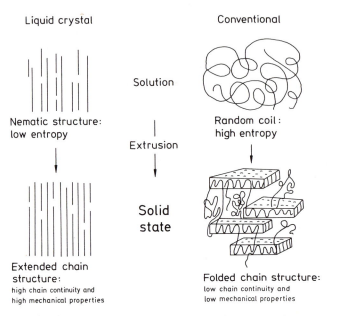

Fig. 2.32. Comparison of dry jet–wet spinning of nematic liquid crystalline solution and conventional spinning of a polymer. (Reprinted from Ref. 51, p. 349, by courtesy of Marcel Dekker, Inc.)

trated H_2SO_4) and extruded through spinnerets at about 100°C through about a 1-cm air layer into cold water (0–4°C) [50]. The fiber precipitates in the air gap and the acid is removed in the coagulation bath. The spinneret capillary and air gap cause rotation and alignment of the domains resulting in highly crystalline and oriented as-spun fibers [49]. Figure 2.32 compares the dry jet–wet spinning method used with nematic liquid crystals and the spinning of a conventional polymer. The oriented chain structure together with molecular extension is achieved with dry jet–wet spinning. The conventional wet or dry spinning gives precursors that need further processing for a marked improvement in properties [51]. The as-spun fibers are washed in water, wound on a bobbin, and dried. Fiber properties are modified by the use of appropriate solvent additives, by changing the spinning conditions, and by means of some postspinning heat treatments.

Structure

Chemically, the Kevlar fiber is poly (*p*-phenyleneterephthalamide), which is a polycondensation product of terephthaloyl chloride and *p*-phenylene diamine. Its chemical formula is given in Fig. 2.33.

The aromatic rings impart the rigid rodlike characteristics of Kevlar. These chains are highly oriented and extended along the fiber axis with the resultant high longitudinal modulus. Kevlar has a highly crystalline structure and the linearity of the polymer chains results in a high packing efficiency. Strong covalent bonding in the fiber direction and weak hydrogen bonding in the transverse direction (Fig. 2.34) result in highly anisotropic properties of Kevlar fiber.

Dobb et al. [52] have examined the structure of Kevlar fiber by electron microscopy and diffraction. They have advanced a schematic representation of the su-

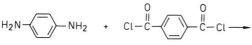

Para-phenylene
diamine

Terephthaloyl
chloride

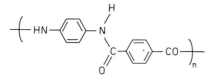

Poly-p-phenyleneterephthalamide (PPDT)

Fig. 2.33. Chemical structure of Kevlar fiber

◀ **Fig. 2.34.** Strong covalent bonding in the fiber direction and weak hydrogen bonding (indicated by H) in the transverse direction

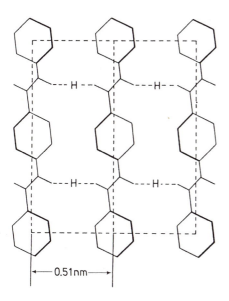

0.51nm

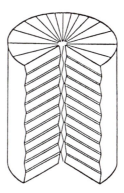

Fig. 2.35. Schematic representation of the supramolecular structure of Kevlar 49. The structure consists of radially arranged, axially pleated crystalline sheets. (From Ref. 52, used with permission.)

pramolecular structure of Kevlar 49 shown in Fig. 2.35. It shows radially arranged, axially pleated crystalline supramolecular sheets. The molecules form a planar array with interchain hydrogen bonding. The stacking sheets form a crystalline array but between the sheets the bonding is rather weak. Each pleat is about 500 nm long and the pleats are separated by transitional bands. The adjacent components of a pleat make an angle of 170°. Such a structure is consistent with the experimentally observed rather low longitudinal shear modulus and poor properties in compression and transverse to the Kevlar fiber axis. A correlation between good compressive characteristics and a high glass transition temperature (or melting point) has been suggested [53]. Thus, the glass transition temperature of organic fibers being lower than that of inorganic fibers, the former would be expected to show poorer properties in compression. For Kevlar and similar fibers, compression results in the formation of kink bands leading to an eventual ductile failure. Yielding is observed at about 0.5% strain. This is thought to correspond to a

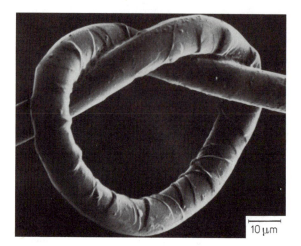

Fig. 2.36. Knotted Kevlar fiber showing buckling marks on the compressive side. The tensile side is smooth. (Courtesy of Fabric Research Corp.)

10 μm

molecular rotation of the amide carbon–nitrogen bond shown in Fig. 2.33 from "the normal extended transconfiguration to a kinked cis configuration" [49]. This causes a 45° bend in the chain. This bend propagates across the unit cell, the microfibrils, and a kink band results in the fiber. This anisotropic behavior of Kevlar is revealed in the SEM micrograph shown in Fig. 2.36. Figure 2.36 shows buckling or kink marks on the compressive side of a knotted Kevlar fiber. Note the absence of such markings on the tensile side. Such markings on the Kevlar fiber surface have also been reported, among others, by DeTeresa et al. [54] when the Kevlar is subjected to uniform compression or torsion.

Properties and Applications

Kevlar aramid fibers provide an impressive array of properties and applications. The fibers are available in three types, each type meant for specific applications [47]:[1]

1. Kevlar — meant mainly for use as rubber reinforcement for tires (belts or radial tires for cars and carcasses of radial tires for trucks) and, in general, for mechanical rubber goods.
2. Kevlar 29 — meant for ropes, cables, coated fabrics for inflatables, architectural fabrics, and ballistic protection fabrics. Vests made of Kevlar 29 have been used by law enforcement agencies in many countries.
3. Kevlar 49 — meant for reinforcement of epoxy, polyester, and other resins for use in aerospace, marine, automotive, and sports industries.

The properties of Kevlar and Kevlar 29 are about the same, the former having a better durability. The main differences are between Kevlar 29 and Kevlar 49. Some

---

[1] *Note added in proof*: A new high modulus variety of Kevlar aramid, called Kevlar 149, was announced by DuPont in 1987. Kevlar 149 is more crystalline than Kevlar 49 and can have a Young's modulus as much as 35% higher than Kevlar 49. Additionally, Kevlar 149 has a moisture regain of about 1/3 that of Kevlar 49, presumably because of a relatively more compact structure of Kevlar 149 (J.E. Van Trump and J. Lahijani, Proc.: 32nd Intl. SAMPE Symposium, April 6–9, 1987, p. 917).

Table 2.5. Properties of Kevlar fibers

	Kevlar 29	Kevlar 49
Density (g cm^{-3})	1.44	1.44
Diameter (μm)	12	12
Tensile strength (GPa)	2.8	2.8
Strain to fracture (%)	4.0	2.3
Young's modulus in tension (GPa)	65	125

of the important properties of these two fibers are summarized in Table 2.5. It must be pointed out that Kevlar fiber has rather poor characteristics in compression, its compressive strength being only about one-eighth of its tensile strength. This follows from the anisotropic nature of the fiber as discussed above. In tensile loading, the load is carried by the strong covalent bonds, while in compressive loading, weak hydrogen bonding and van der Waals bonds come into play, which lead to rather easy local yielding, buckling, and kinking of the fiber. Thus, one would be well advised against using Kevlar fiber in situations involving compressive forces. The solution, in such circumstances, frequently has been to use hybrid composites wherein one mixes Kevlar with, say, carbon fibers in a matrix with carbon taking up the compressive loads and Kevlar aligned parallel to the tensile load paths.

Kevlar fiber being on organic fiber undergoes photodegradation when exposed to light (both visible as well as ultraviolet), which shows up as a discoloration and loss in mechanical properties. One can minimize this problem by adequately coating the Kevlar composite surface with a light-absorbing material.

2.6 Ceramic Fibers

Continuous ceramic fibers present an attractive package of properties. They combine rather high strength and elastic modulus with high-temperature capability and a general freedom from environmental attack. These characteristics make them attractive as reinforcements in high-temperature structural materials.

There are three types of ceramic fiber fabrication method: chemical vapor deposition, polymer pyrolysis, and sol-gel techniques. The latter two routes involve rather novel techniques of obtaining ceramics from organometallic polymers. Mention has been made of the sol-gel technique in Sect. 2.2 regarding the manufacture of glass fibers. The great breakthrough in the ceramic fiber area has been, undoubtedly, the concept of pyrolysing, under controlled conditions, polymers containing silicon and carbon or nitrogen to produce high-temperature ceramic fibers. The idea really can be regarded as an extension of the polymer pyrolysis route to produce a variety of carbon fibers wherein a suitable carbon-based polymer (e.g., PAN and pitch) is subjected to controlled heating to produce carbon fibers (see Sect. 2.4). The pyrolysis route of producing ceramic fibers has been used with polymers containing silicon, carbon, nitrogen, and boron with the end products being SiC,

Si_3N_4, B_4C, and BN in fiber form, foam, or coatings. We describe below some important ceramic fibers.

2.6.1 Alumina Fibers

Alumina fibers of different types are produced by a number of companies. Du Pont produces a continuous filament polycrystalline α-alumina yarn. Sumitomo Chemical Co. produces a fiber that may be composed of 70–100% Al_2O_3 and 30–0% SiO_2. ICI produces short staple, δ-alumina fibers (trade name Saffil). We describe below the salient features of some of these fabrication methods and the properties of fibers obtained.

Fiber FP

Fiber FP is a continuous α-alumina yarn with a 98% theoretical density. Dhingra [55] has described the basic fabrication procedure. The three basic steps involved are as follows:

1. An aqueous slurry mix is made of selected alumina particles and some additives to render it spinnable; the viscosity of this slurry is controlled by controlling the amount of water present.
2. Fibers are dry spun from this spinnable slurry.
3. Finally, the dry spun yarn is subjected to two-step firing: Low firing controls the shrinkage while flame firing improves the density of α-Al_2O_3. A thin silica coating is generally applied. It serves to heal the surface flaws, giving about 50% higher tensile strength than the uncoated fibers.

Figure 2.37 gives the flow diagram of the process, while Table 2.6 presents the

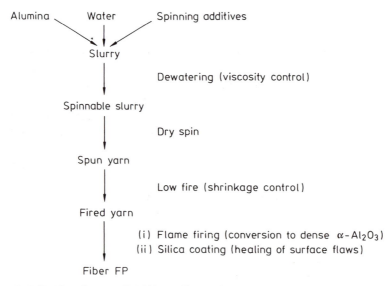

Fig. 2.37. Flow diagram of FP (Al_2O_3) fiber production. (Adapted from Ref. 55, used with permission.)

Table 2.6. Properties of fiber FP[a]

Diameter (μm)	Density (g cm^{-3})	Tensile strength (MPa)	Young's modulus (GPa)	Melting point (°C)
20 ± 5	3.95	1380	379	2045

[a] α-Al$_2$O$_3$.
Source: Adapted with permission from Ref. 55.

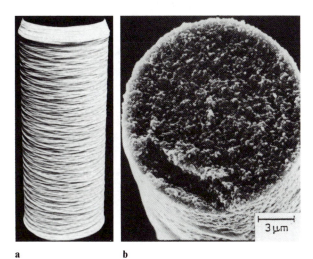

Fig. 2.38. a A bobbin of continuous FP (Al$_2$O$_3$) fiber. (Courtesy of Du Pont Co.) **b** Rough ("cobblestone") surface texture of as-produced FP fiber

a   b

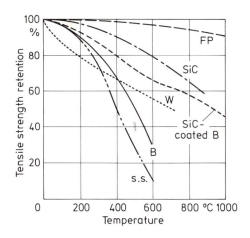

Fig. 2.39. Tensile strength retention at high temperature for some fibers. s.s. is stainless steel. (Adapted from Ref. 55, used with permission.)

properties of FP fiber. Figure 2.38a shows a bobbin of continuous FP fiber, while Fig. 2.38b shows a scanning electron micrograph of an individual FP fiber showing that the as-produced FP fiber surface has a very rough texture. An excellent feature of FP fiber is its strength retention at high temperatures. A comparison of this strength retention for some fibers is made in Fig. 2.39 [55].

Other Alumina Fibers

A variety of other alumina fibers is available. Sumitomo Chemical Co. produces a mixture of alumina and silica fiber. The flow diagram of this process is shown in Fig. 2.40. Basically, the process starts with an organoaluminum (polyaluminoxanes or a mixture of polyaluminoxanes and one or more kinds of silicon-containing compounds) and a precursor fiber is obtained by dry spinning. This precursor fiber is calcined to produce the final fiber. 3M Company produces a ceramic fiber having 62% Al_2O_3 14% B_2O_3, and 24% SiO_2. The fiber has the trade name Nextel 312. The manufacturing process involves a sol-gel technique and starting materials are solutions of metal alkoxides. Metal alkoxides are $M(OR)n$ type compounds where M is the metal, n is the metal valence, and R is an organic compound. Selection of an appropriate organic group is very important. It should provide sufficient stability and volatility to the alkoxide so that M—OR bond is broken and MO—R is obtained to give the desired oxide ceramics [56]. Hydrolysis of metal alkoxides results in sols that are spun and gelled. The gelled fiber is then densified at intermediate temperatures. The high surface free energy available in the pores of the

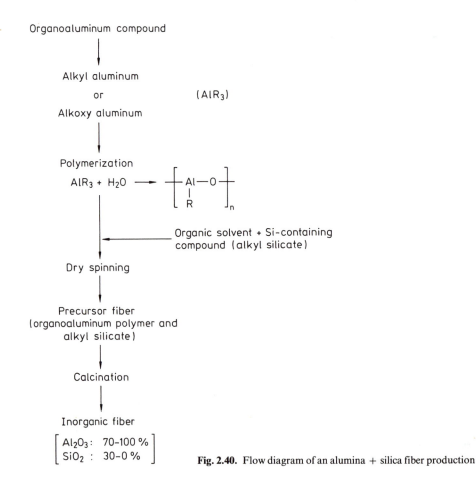

Fig. 2.40. Flow diagram of an alumina + silica fiber production

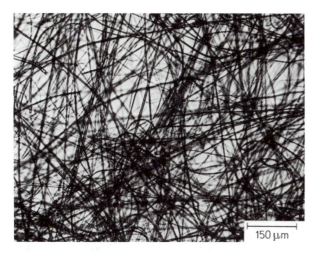

150 μm

Fig. 2.41. Optical micrograph of Nextel 312 (Al_2O_3 + B_2O_3 + SiO_2) fiber

Table 2.7. Properties of some alumina type fibers

Fiber type	Composition	Diameter (μm)	Density (g cm^{-3})	Tensile strength (MPa)	Young's modulus (GPa)
Sumitomo[a]	Al_2O_3 — 85% SiO_2 — 15%	9	3.2	2600	250
Nextel 312 (3M Co.)	Al_2O_3 — 62% B_2O_3 — 14% SiO_2 — 24%	3.5	2.7	1700	152
Saffil (ICI)	Al_2O_3 — 96% SiO_2 — 4%	3	3.3	2000	300

[a] Laboratory scale production.

gelled fiber allows for a relatively low-temperature densification. Figure 2.41 shows an optical micrograph of Nextel fibers. Yet another variety of alumina fiber available commercially is δ-Al_2O_3, a short staple fiber produced by ICI (trade name Saffil). This fiber has about 4% SiO_2 and a very fine diameter (3 μm). It is also produced by spinning and calcining at high temperature. Table 2.7 summarizes the properties of these three fibers.

2.6.2 Silicon Carbide Fibers

We can easily classify the fabrication methods for SiC as conventional and non-conventional. The former category would include the chemical vapor deposition process, while the latter would include controlled pyrolysis of polymeric precursors.

There is yet another important type of SiC available for reinforcement purposes, namely, SiC whiskers. We give a brief description of these.

Chemical Vapor Deposition [57]

SiC is chemically vapor deposited on a tungsten (sometimes carbon) substrate heated to around 1300°C. The reactive gaseous mixture contains hydrogen and alkyl silanes. Typically, a gaseous mixture consisting of 70% hydrogen and 30% silanes is introduced at the reactor top (Fig. 2.42), where the tungsten substrate (~ 13 μm diameter) also enters the reactor. Mercury seals are used at both ends as contact electrodes for the filament. The substrate is heated by a combined direct current (250 mA) and a very high frequency (VHF ≈ 60 MHz) to obtain an optimum temperature profile. To obtain a 100-μm SiC monofilament, it generally takes about 20 s in the reactor. The filament is wound on a spool at the bottom of the reactor. The exhaust gases (95% of the original mixture + HCl) are passed around a condenser to recover the unused silanes. An efficient reclamation of the unused silanes is very important for a cost effective production process. The reader will have noted the similarities between the CVD processes for SiC and B fibers. Such CVD processes result in composite monofilaments which have built-in residual stresses. The process, of course, is very expensive. Methyltrichlorosilane is an ideal raw material because it contains one silicon atom and one carbon atom; that is, one would expect a stoichiometric SiC to be deposited. The chemical reaction is

$$CH_3SiCl_3 \rightarrow SiC + 3HCl$$

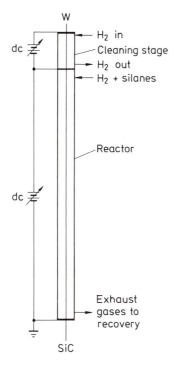

Fig. 2.42. CVD process for SiC monofilament fabrication. (Reproduced by permission of the University of South Carolina Press from *Production of Silicon Carbide from Rice Hulls* by J.V. Milewski, J.L. Sandstrom, and W.S. Brown in *Silicon Carbide-1973* edited by R.C. Marshall, J.W. Faust, Jr., and C.E. Ryan.)

Table 2.8. Properties of CVD SiC monofilament

Composition	Diameter	Density	Tensile strength	Young's modulus
	(µm)	(g cm⁻³)	(MPa)	(GPa)
β-SiC	140	3.3	3500	430

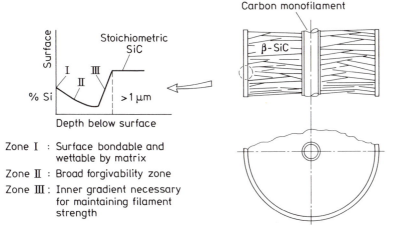

Zone I : Surface bondable and wettable by matrix

Zone II : Broad forgivability zone

Zone III : Inner gradient necessary for maintaining filament strength

Fig. 2.43. Schematic of AVCO SCS-6 silicon carbide fiber and its surface compositional gradient. (Courtesy of AVCO Specialty Materials Co.)

Generally, solid (free) carbon and solid or liquid silicon are mixed with SiC. The final monofilament (100–150 µm) consists of a sheath of mainly β-SiC with some α-SiC on the tungsten core. The cross section of SiC monofilament resembles very much that of a boron fiber. Properties of a CVD SiC monofilament are given in Table 2.8.

AVCO Specialty Materials Co. has developed a silicon carbide fiber, called SCS fiber, that is eminently suitable for reinforcement of metals. SCS-2 is meant for aluminum and magnesium while SCS-6 is meant for titanium. These special fibers have a complex through the thickness gradient structure. SCS-6, for example, is a thick fiber (diameter = 140 µm) and is produced by CVD of silicon- and carbon-containing compounds onto a pyrolytic graphite-coated carbon core. The bulk of the ~ 1-µm thick surface coating consists of carbon-doped silicon. Figure 2.43 shows schematically the AVCO SCS silicon carbide fiber and its characteristic surface composition gradient. Zone I at and near the surface is a zone of easy wetting and bonding with the matrix. This zone I is carbon rich. This sacrificial layer serves to promote bonding and may very well react with the matrix without degrading the fiber. Zone II, wherein the silicon content decreases somewhat, has been called the broad forgivability zone. This is followed by a zone III in which the silicon content increases back to stoichiometric SiC composition. In SCS, silicon carbide fiber has a surface graded outward to be carbon rich and back to stoichiometric SiC at a few micrometers from the surface. This fiber gives a *buffer* layer at the surface that allows fiber strength to be maintained even during the high-temperature incorporation into a metal matrix.

Ceramic Fibers via Polymers

As pointed out above, the SiC fiber obtained via CVD is very thick and not very flexible. Work on alternate routes of obtaining fine, continuous, and flexible fiber had been going on for some time when, in the mid-1970s, the late Professor Yajima and his colleagues [58, 59] in Japan developed a process of making such a fiber by controlled pyrolysis of polymeric precursor. Not surprisingly, Japan has assumed a position of leadership in producing continuous SiC fiber under the trade name Nicalon. A number of review articles summarizing the state of the art regarding fabrication and properties of SiC fiber until about the mid-1980s is available [60–62]. This method of using silicon-based polymers to produce a family of ceramic fibers having good mechanical properties, good thermal stability, and oxidation resistance has an enormous potential. The various steps involved in this polymer route, shown in Fig. 2.44 [63], are:

1. Identify a suitable starting polymer.
2. Devise an efficient polymer preparation method.
3. Characterize the polymer (e.g., yield, molecular weight, and purity).
4. Melt spin the polymer into a precursor fiber.
5. Cure the precursor fiber to crosslink the molecular chains, making it infusible during the subsequent pyrolysis.
6. Pyrolyze to obtain the ceramic fiber.

Specifically, the Yajima process of making SiC involves the following steps and is

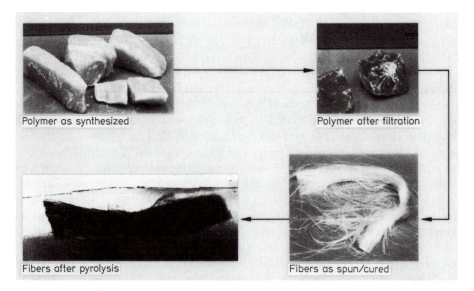

Polymer as synthesized

Polymer after filtration

Fibers after pyrolysis

Fibers as spun/cured

Fig. 2.44. Schematic of ceramic fiber production starting from silicon-based polymers. (Adapted from Ref. 63, used with permission.)

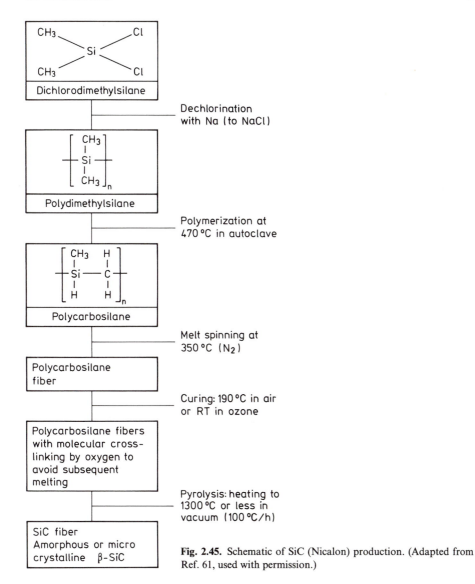

Fig. 2.45. Schematic of SiC (Nicalon) production. (Adapted from Ref. 61, used with permission.)

shown schematically in Fig. 2.45. Polycarbosilane, a high molecular weight polymer containing silicon and carbon, is synthesized. The starting material is commercially available dichlorodimethylsilane. Solid polydimethylsilane is obtained by dechlorination of dichlorodimethylsilane by reacting it with sodium. Polycarbosilane is obtained by thermal decomposition and polymerization of polydimethylsilane. This is carried out under high pressure in an autoclave at 470°C in an argon atmosphere for 8–14 h. A vacuum distillation treatment at up to 280°C follows. The average molecular weight of the resulting polymer is about 1500. This is melt spun from a 500 hole nozzle at about 350°C under N_2 gas to obtain the so-called preceramic continuous, precursor fiber. The precursor fiber is quite weak (tensile strength $\simeq 10$ MPa). This is converted to inorganic SiC by curing in air, heating to about 1000°C in N_2 gas, followed by heating to 1300°C in N_2 under stretch. This basically is the Nippon Carbon Co. (NCK) manufacture process for Nicalon fibers [64]. Clearly, small variations exist between this and the various laboratory processes. During this pyrolysis, the first stage of conversion occurs at around 550°C when cross linking of polymer chains occurs. Above this temperature, the sidechains containing hydrogen and methyl groups decompose. Fiber density and mechanical properties improve sharply. The conversion to SiC is complete above about 850°C.

Structure and Properties

A number of workers has studied the structure of Nicalon fibers. Figure 2.46 shows a high-resolution transmission electron micrograph of Nicalon type SiC produced in the laboratory, indicating the amorphous nature of the SiC produced by the Yajima process. Simon and Bunsell [65] found the commercial variety of Nicalon to have an amorphous structure while another, noncommercial variety showed a microcrystalline structure (SiC grain radius of 1.7 nm). Their microstructural

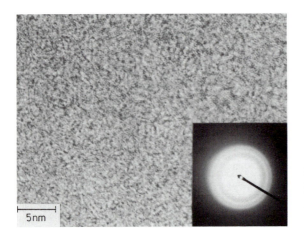

Fig. 2.46. High-resolution transmission electron micrograph of Nicalon fiber showing its amorphous structure. (Courtesy of K. Okamura.)

Table 2.9. Typical properties of Nicalon SiC fiber

Density	2.6 g cm^{-3}
Diameter	10–20 μm (500 per yarn)
Modulus	180 GPa (420 GPa for β-SiC)
Strength at 20°C	
As-produced	2 GPa
After 1400°C	<1 GPa
(argon)	
Strength at 1400°C	<0.5 GPa
(oxygen)	
Creep strain	
at 1300°C, 0.6 GPa, 20h	4.5%

Source: Reprinted with permission from *Journal of Metals,* **37**, No. 6, 1985, a publication of The Metallurgical Society, Warrendale, Pennsylvania.

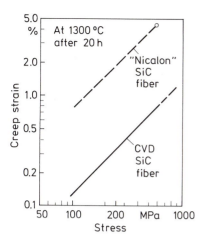

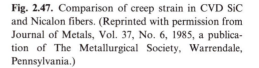

Fig. 2.47. Comparison of creep strain in CVD SiC and Nicalon fibers. (Reprinted with permission from Journal of Metals, Vol. 37, No. 6, 1985, a publication of The Metallurgical Society, Warrendale, Pennsylvania.)

analysis showed that both the fibers contained, in addition to SiC, SiO_2 and free carbon. The density of the fiber is about $2.6\,\mathrm{g\,cm^{-3}}$ which is low compared to that of pure β-SiC; this is not surprising in view of the fact that the composition is a mixture of SiC, SiO_2, and C.

The properties of Nicalon fiber are summarized in Table 2.9. A comparison of Nicalon SiC fiber with CVD SiC fiber, produced by AVCO (see Table 2.8), shows that CVD fiber is superior in properties mainly because it is almost 100% β-SiC while the Nicalon fiber is a mixture of SiC, SiO_2, and free carbon. Figure 2.47 shows a comparison of the creep strain in Nicalon and CVD SiC fiber. Note the superior performance of the CVD fiber. The CVD fiber, it should be pointed out, is a lot more expensive than the Nicalon fiber.

SiC Whiskers and Particles

SiC in particulate form has been available quite cheaply and abundantly for abrasive, refractory, and chemical uses. In this conventional process, silica in the form of sand and carbon in the form of coke are made to react at 2400°C in an electric furnace. The SiC produced in the form of large granules is subsequently comminuted to the desired size.

Whiskers are normally obtained by vapor phase growth. They are almost monocrystalline, short fibers with extremely high strength. Typically, whiskers have a diameter of a few micrometers and a length of a few millimeters. Thus, their aspect ratio (length/diameter) can vary from 50 to 10,000. Whiskers do not have uniform dimensions or properties. This is perhaps their greatest disadvantage; that is, the spread in properties is extremely large. Handling and alignment of whiskers in a matrix to produce a composite are other problems. Early in the 1970s, a new process was developed, starting from rice hulls, to produce SiC particles and whiskers [66, 67]. The SiC particles produced by this process are of a finer size. Rice hulls are a waste by-product of rice milling. For each 100 kg of rice milled, about 20 kg of rice hull are produced. Rice hulls contain cellulose, silica, and other organic and inorganic materials. Silica from soil is dissolved and transported in the plant as monosilicic acid. This is deposited in the cellulosic structure by liquid evaporation. It turns out that most of the silica ends up in the hull. It is the intimate mixture of silica within the cellulose that gives the near ideal amounts of silica and carbon for silicon carbide production. Raw rice hulls are heated in the absence of oxygen at about 700°C to drive out the volatile compounds. This is called coking. Coked rice hulls, containing about equal amounts of SiO_2 and free carbon, are heated in an inert or reducing atmosphere (flowing N_2 or NH_3 gas) at a temperature between 1500 and 1600°C for about 1 h to form silicon carbide as per the following reaction:

$$3\,C + SiO_2 \rightarrow SiC + 2\,CO$$

Figure 2.48a shows the schematic of the process, while Fig. 2.48b shows a scanning electron micrograph of these whiskers. When the above reaction is over, the residue

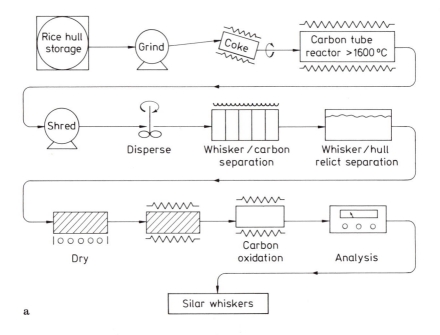

a

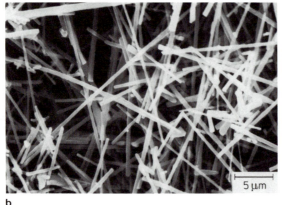

b

Fig. 2.48. a Schematic of SiC whisker production process starting from rice hulls. b Scanning electron micrograph of SiC whiskers obtained from rice hulls. (Courtesy of ARCO Chemical Co.)

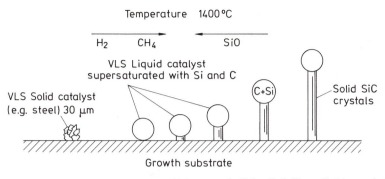

Fig. 2.49. The VLS process for SiC whisker growth. (After Ref. 68, used with permission.)

is heated to 800°C to remove any free carbon. Generally, both particles and whiskers are produced together with some excess free carbon. A wet process is used to separate the particles and the whiskers. Typically, the average aspect ratio of the as-produced whiskers is 50.

Exceptionally strong and stiff silicon carbide whiskers have been grown at the Los Alamos National Laboratory (LANL) [68, 69]. Their average tensile strength and modulus were 8.4 GPa and 581 GPa, respectively. Compared to Nicalon SiC fiber (2.4 GPa and 180 GPa, respectively), these values are indeed very high. The LANL whiskers were grown by the VLS process. The acronym VLS stands for vapor feed gases, liquid catalyst, and solid crystalline whiskers. The catalyst forms a liquid solution interface with the growing crystalline phase while elements are fed from the vapor phase through the liquid–vapor interface. Whisker growth takes place by precipitation from the supersaturated liquid at the solid–liquid interface. The catalyst must take in solution the atomic species of the whisker to be grown. For SiC whiskers, transition metals and iron alloys meet this requirement. Milewski et al. [68] give details of their process of growing β-SiC whiskers. Figure 2.49 illustrates the LANL process for SiC whisker growth. Steel particles (~ 30 μm) are used as catalyst. At 1400°C, the solid steel catalyst particle melts and forms a liquid catalyst ball. From the vapor feed of SiO, H_2, and CH_4, the liquid catalyst extracts C and Si atoms and forms a supersaturated solution. The gaseous silicon monoxide is generated as per the following reaction

$$SiO_2(g) + C(s) \rightarrow SiO(g) + CO(g)$$

The supersaturated solution of C and Si in the liquid catalyst precipitates out solid SiC whisker on the substrate. As the precipitation continues with time, the whisker grows (Fig. 2.49). They have identified a range of whisker morphologies. The tensile strength values ranged from 1.7 to 23.7 GPa in 40 tests. Whisker lengths were about 10 mm and the equivalent circular diameter averaged 5.9 μm. It must be borne in mind, however, that the LANL results pertain to a laboratory scale product, while Nicalon is a full fledged commercial product and a continuous fiber.

Other Ceramic Fibers

Besides alumina and silicon carbide ceramic fibers, there are other promising ceramic fibers, for example, silicon nitride, boron carbide, and boron nitride.

Silicon nitride (Si_3N_4) fibers can be prepared by reactive CVD using volatile silicon compounds. The reactants are generally $SiCl_4$ and NH_3. Si_3N_4 is deposited on a carbon or tungsten substrate. Again, as in other CVD processes, the resultant fiber has good properties but the diameter is very large and is expensive. In the polymer route, organosilazane[1] polymers with methyl groups on silicon and nitrogen have been used as silicon nitride precursors. Such carbon-containing, silicon–nitrogen precursors on pyrolysis give silicon carbide as well as silicon nitride; that is, the resulting fiber is not SiC-free silicon nitride fiber. Wills et al. [70] have discussed the mechanisms involved in the coversion of various organometallic compounds into ceramics.

Boron nitride is very competitive with carbon fiber. It has about the same density (2.2 g cm^{-3}) as the carbon fiber but has a greater oxidation resistance and excellent dielectric properties. A method of converting boric oxide precursor fibers into boron nitride fibers has been developed [71, 72]. Melt spun boric oxide precursor fiber is nitrided with ammonia as per the following reaction:

$$B_2O_3 + 2NH_3 \rightarrow 2BN + 3H_2O$$

This is followed by a high-temperature treatment that removes the traces of oxides and stabilizes the product [71].

Boron carbide is also a very light and strong material. It can be prepared by reacting carbon yarn with BCl_3 and H_2 at high temperatures, that is, a CVD process [73]. The chemical reaction involved is

$$4BCl_3 + 6H_2 + C_{fibers} \rightarrow B_4C_{fibers} + 12HCl$$

The reaction actually occurs in two steps, namely,

$$2BCl_3 + 3H_2 \rightarrow 2B + 6HCl$$

$$4B + C \rightarrow B_4C$$

Much like other CVD processes described earlier, the gaseous mixture of BCl_3, H_2, and argon (diluent) enters at one end of a furnace, reacts in the hot zone, and the reaction products exit at the other end. The second step in the above reaction is the rate controlling step because the reaction of boron and carbon is slowed by the formation of a B_4C layer.

2.7 Metallic Fibers

Many metals in the form of wires show rather high strength levels. The wires of beryllium (low density and high modulus), steel (high strength and low cost), and tungsten (high modulus and refractory) are among the most important ones. One great advantage of metallic wires is that they show very consistent strength values,

---

[1] A silazane compound has Si-NH-Si bonds.

more so than any of the ceramic fibers. Beryllium, in particular, with its high modulus (~ 300 GPa) and extremely low density (1.8 g cm^{-3}) would make for an ideal reinforcement candidate but for its toxicity and high cost. Its strength is relatively low (~ 1300 MPa).

Tungsten wires were originally developed for lamps. Tungsten has a very high density (19.3 g cm^{-3}) but because of its refractory properties, it has been used as a reinforcement in some nickel- and cobalt-based superalloys [74], in addition to the tungsten fiber–copper matrix system which has been extensively studied as a model fiber reinforcement system [75–78]. Some applications of tungsten–copper composites as electrical contacts have been reported [79]. One of the problems of tungsten is its ease of oxidation and the oxide of tungsten is likely to volatilize at high service temperatures.

Steel wire is a common commercial reinforcement material, more for concrete than for metals or polymers. Steel wires are also commonly used as reinforcement in tires. Very fine (0.1 mm diameter) and high carbon (0.9%) steel wires can have very high strength levels (~ 5 GPa), although the toughness levels will be rather low at such high strengths. Production of fine metallic wires is very expensive owing to the cost of the wire drawing process (the Taylor process). It turns out, for example, that for steel wires with diameters less than 25 μm, the cost of producing wire becomes constant per unit length and not per unit weight; that is, the material cost at such fine diameters is not very high but the processing costs are.

Coventional wire drawing methods are quite reasonable for producing wires of titanium, tungsten, tantalum, molybdenum, steels, and so on with diameters down to 100 μm. The production costs increase tremendously below this diameter. Wires of diameters down to 10 μm or less can be obtained by the so-called Taylor process. The metallic wire is encased in a sheath of a sacrificial material (e.g., glass), heated to a temperature where the sheath becomes quite soft and the core wire melts or softens, followed by drawing the whole thing in plastic state down to very fine diameter and removing the sheath material by etching. Mechanical properties of some commercial metallic wires are summarized in Table 2.10.

Table 2.10. Typical properties of some commercial metallic wires[a]

Material	Diameter	Density	Tensile strength	Young's modulus	Coefficient of thermal expansion	Melting point
	(μm)	(g cm^{-3})	(MPa)	(GPa)	(10^{-6} K^{-1})	(°C)
Steel, 0.9%C	100	7.8	4250	210	11.8	1300
Stainless steel, 18-8	50–250	8.0	700–1000	198	18.0	—
Beryllium	—	1.85	1260	300	11.6	1280
Tungsten	<25	19.3	3850	360	4.5	3400
	125		3150			
	250					
Molybdenum	<25	10.2	2450	310	6.0	2600

[a] These values are indicative only. Heat treatments (e.g., annealing and aging) and processing can alter the strength properties quite drastically.

2.7.1 Metallic Glasses

Metals are crystalline materials. Under extreme processing conditions, however, it is possible to obtain metals with a glassy structure. We call these amorphous metals or metallic glasses. Metallic glasses are obtained by ultrarapid quenching of metallic melts. Cooling rates in the range of 10^5-10^8 K s^{-1} are required. By contrast, the conventional oxide glasses require cooling rates of a few tens of degrees per second. The field of metallic glasses saw rapid advances in the 1960s and 1970s and a few companies, notably the Allied Corporation, produce commercially available metallic glass ribbons. In most processes, these ribbons are produced by some variant of melt spinning involving a thin layer of molten alloy of suitable composition coming in contact with a medium of excellent heat conductivity (usually copper). Ribbons several kilometers long, 25–100 μm thick, and 1–15 mm wide can be made at speeds of 5–20 m s^{-1}.

Although metallic glasses are structurally similar to oxide glasses, that is, they are amorphous, there are some important differences. Metallic glasses consist primarily of metallic elements; they are opaque and have mainly metallic bonding. Their mechanical properties are close to or superior to those of conventional metals. Most metallic alloy compositions are binary, ternary, or more complex alloy systems. Typically, metallic glasses have densities about 2% less and shear and Young's moduli about 20–30% lower than those of their crystalline counterparts [80]. Their fracture toughness values are about two orders of magnitude greater than those of oxide glasses [80]. Some typical metallic glass compositions as well as some of their mechanical properties are presented in Table 2.11. There exists an extensive bibliography on the subject and the reader is referred to Refs. [80–83] for greater details.

Table 2.11. Mechanical properties of metallic glasses compared with those of commercial steels

Alloy metallic glasses	VH[a] (kg mm^{-2})	Yield strength (MPa)	Density (g cm^{-3})
$Pd_{77.5}Cu_6Si_{16.5}$	500	1600	10.30
$Zr_{50}Cu_{50}$	580	1800	7.33
$Ti_{60}Be_{35}Si_5$	805	2530	3.90
$Fe_{80}P_{16}C_3B_1$	835	2500	7.30
$Fe_{80}B_{20}$	1100	3700	7.40
$Co_{60}W_{30}B_{10}$	1600	3600	—
$Fe_{60}Cr_6Mo_6B_{28}$	—	4900	—
AISI 4340[b] (steel)	450	1700	—
18% Ni (ausformed)[c] (steel)	600	2050	8.0

[a] VH, Vickers hardness.
[b] 0.42 C, 0.78 Mn, 1.79 Ni, 0.8 Cr, 0.33 Mo.
[c] 18 Ni, 8 Co, 3 Mo, 0.2 Ti, 0.1 A8.
Source: Adapted with permission from Ref. 80.

2.8 Comparison of Fibers

A comparison of some important characteristics of high-performance reinforcements discussed individually in Sect. 2.2–2.7 is made in Table 2.12 and a plot of strength versus modulus is shown in Fig. 2.50. We compare and contrast below some salient points of these fibers.

First, we note that all these so-called advanced fibers have very low density values. Given the general low density of these fibers, the best of these advanced fibers group together in the top right-hand corner of Fig. 2.49. Second, the reader will also recognize that the elements comprising these fibers pertain to the first two rows of the Periodic Table. Third, irrespective of whether the elements are in compound or elemental form, they are mostly covalently bonded which is the strongest bond. Fourth, such light, strong, and stiff materials are very desirable in most applications, but particularly so in the aerospace field, land transportation, energy-related industries, and housing and civil construction.

Fiber flexibility is associated with Young's modulus and the diameter (see Sect. 2.1). In the general area of high-moduli fibers, the diameter becomes the dominant parameter controlling the flexibility. For a given E, the smaller the diameter the more flexible it is. Fiber flexibility allows it to be woven, bent, and wound into intricate shapes. Here, the difficulty lies with monofilament type fibers produced by CVD techniques (e.g., boron and SiC).

Some of these fibers have quite anisotropic characteristics. Kevlar, in particular, has rather poor strength properties in compression compared to those in tension. This is due to its structure as explained in Sect. 2.5. Covalent bonds carry the load in tension but in compression weak hydrogen bonding comes into play, leading to local yielding, buckling, and kink formation. Other fibers show quite similar properties in tension and compression. The situation is different with regard to thermal properties; in particular, the thermal expansion coefficients of carbon and Kevlar fibers are quite different in the radial and longitudinal directions. Besides this anisotropy in expansion, there is the question of amount of retention of ambient level properties at elevated temperatures. It is clear that organic fibers, such as Kevlar and UHMW polyethylene may only be used with polymeric matrices and only at or about room temperature. In this respect, only ceramic (SiC and Al_2O_3), carbon, and boron fibers are viable in metal or ceramic matrices. Ceramic matrix composites can go to very high temperatures indeed. A problem that arises at these very high temperatures ($> 1500°C$) is that of fiber and matrix oxidation. Carbon fiber, for example, does not have a great oxidation resistance at these high temperatures. SiC or Si_3N_4 type ceramic fibers are the only suitable candidates for reinforcement at very high temperatures ($> 1200–1300°C$) and in air.

Another important characteristic of these high-performance fibers is their rather low values of strain to fracture, that is, generally less than 2–3%. This rather low ductility coupled with their slenderness makes it almost mandatory that these fibers should be distributed in a binding medium called a matrix. This would allow us to make useful components out of the resultant composite materials. The function of the matrix, however, is not just that of holding the fibers together. It also serves to transmit the applied load to the fibers, generally the major load-bearing component.

Table 2.12. Properties of reinforcement fibers

Characteristic	PAN-based carbon		Kevlar 49	E glass	SiC		Al$_2$O$_3$(FP)	Boron (W)
	HM	HS			CVD	Nicalon		
Diameter (μm)	7–10	7.6–8.6	12	8–14	100–200	10–20	20	100–200
Density (g cm^{-3})	1.95	1.75	1.45	2.55	3.3	2.6	3.95	2.6
Young's modulus (GPa)								
Parallel to fiber axis	390	250	125	70	430	180	379	385
Perpendicular to fiber axis	12	20	—	70	—	—	—	—
Tensile strength (GPa)	2.2	2.7	2.8–3.5	1.5–2.5	3.5	2	1.4	3.8
Strain to fracture (%)	0.5	1.0	2.2–2.8	1.8–3.2	—	—	—	—
Coefficient of thermal expansion (10^{-6} K^{-1})								
Parallel to fiber axis	−0.5 to 0.1	0.1 to −0.5	−2 to −5	4.7	5.7	—	7.5	8.3
Perpendicular to fiber axis	7–12	7–12	59	4.7	—	—	—	—

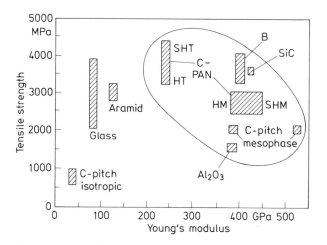

Fig. 2.50. Comparison of different fibers

It may also contribute its own distinctive characteristics, say high fracture toughness, and thus make a more crack-resistant composite. We should add that the nature of interfacial bonding between fiber and matrix can also influence greatly the strength and toughness of a composite. We describe in the next two chapters some characteristics of matrix materials (Chap. 3) and the interface (Chap. 4).

References

1 K.K. Chawla, in *Proceedings of the International Conference on the Mechanical Behavior of Materials II*, ASM, Metals Park, Ohio, 1976, p. 1920.
2 K.K. Chawla and A.C. Bastos, in *Proceedings of the International Conference on the Mechanical Behavior of Materials III*, Pergamon Press, Oxford, 1979, p. 191.
3 R. Meredith, *Contemp. Phys.* **11**, 43 (1970).
4 W.H. Dresher, *J. Metals*, **21**, 17 (Apr. 1969).
5 E. de Lamotte and A.J. Perry, *Fibre Sci. Tech.*, **3**, 157 (1970).
6 K.L. Loewenstein, *The Manufacturing Technology of Continuous Glass Fibers*, 2nd ed., Elsevier, New York, 1983.
7 B. Parkyn (ed.), *Glass Reinforced Plastics*, Butterworth, London, 1970.
8 R.E. Lowrie, in *Modern Composite Materials*, Addison-Wesley, Reading, MA, 1967, p. 270.
9 C.J. Brinker, D.E. Clark, and D R. Ulrich (eds.), *Better Ceramics Through Chemistry*, North-Holland, New York, 1984.
10 T. Davis, H. Palmour, and T. Porter (eds.), *Emergent Process Methods for High Technology Ceramics*, Plenum Press, New York, 1984.
11 S. Sakka, *Am. Ceram. Soc. Bull.*, **64**, 1463 (1985).
12 E. Weintraub, *J. Ind. Eng. Chem.*, **3**, 299 (1911).
13 C.P. Talley, *J. Appl. Phys.*, **30**, 1114 (1959).
14 C.P. Talley, L. Line, and O. Overman, in *Boron: Synthesis*, Structure, *and Properties*, Plenum Press, New York, 1960, p. 94.
15 A.C. van Maaren, O. Schob, and W. Westerveld, *Philips Tech. Rev.*, **35**, 125 (1975).
16 V. Krukonis, in *Boron and Refractory Borides*, Springer-Verlag, Berlin, 1977, p. 517.

17   J. Vega-Boggio and O. Vingsbo, in *1978 International Conference on Composite Materials, ICCM/2*, TMS-AIME, New York, 1978, p. 909.

18   F. Galasso, D. Knebl, and W. Tice, *J. Appl. Phys.*, **38**, 414 (1967).

19   F. Galasso and A. Paton, *Trans. Met. Soc. AIME*, **236**, 1751 (1966).

20   H.E. DeBolt, in *Handbook of Composites*, Van Nostrand Reinhold, New York, 1982, p. 171.

21   F.W. Wawner, in *Modern Composite Materials*, Addison-Wesley, Reading, MA, 1967, p. 244.

22   J.A. DiCarlo, *J. Met.* **37**, 44 (June 1985).

23   K.K. Chawla, *Mater. Sci. Eng.*, **48**, 137 (1981).

24   A. Shindo, *Rep. Osaka Ind. Res. Inst.* No. 317 (1961).

25   A.A. Baker, *Metals Forum*, **6**, 81 (1983).

26   W. Watt, *Proc. R. Soc.*, **A319**, 5 (1970).

27   R. Bacon, in *Chemistry and Physics of Carbon*, vol. 9, Marcel Dekker, New York, 1973, p. 1.

28   R.J. Diefendorf and E. Tokarsky, *Polym. Eng. Sci.*, **15**, 150 (1975).

29   H.N. Ezekiel and R.G. Spain, *J. Polym. Sci. C*, **19**, 271 (1967).

30   W. Watt and W. Johnson, *Appl. Polym. Symp.*, **9**, 215 (1969).

31   D.J. Johnson and C.N. Tyson, *Br. J. Appl. Phys.*, **2**, 787 (1969).

32   R. Perret and W. Ruland, *J. Appl. Crystallogr.*, **3** (1970) 525.

33   S.C. Bennett and D.J. Johnson, in *5th International Carbon and Graphite Conference*, Society of the Chemical Industry, London, 1978, p. 377.

34   S.C. Bennett and D.J. Johnson, *Carbon*, **17**, 25 (1979).

35   O.T. Inal, N. Leca, and L. Keller, *Phys. Status Solidi*, **62**, 681 (1980).

36   S.C. Bennet, D.J. Johnson, and W. Johnson, *J. Mater. Sc.*, **18**, 3337 (1983), Chapmann & Hall.

37   A. Fourdeux, R. Perret, and W. Ruland, in *Carbon Fibres: Their Composites and Applications*, The Plastics Institute, London, 1971, p. 57.

38   L.S.Singer, in *Ultra-High Modulus Polymers*, Applied Science Publishers, Essex, England, 1979, p. 251.

39   J.P. Riggs, in *Encyclopedia of Polymer Science & Engineering*, 2nd ed., vol. 2, John Wiley & Sons, New York, 1985, p. 640.

40   J.S. Murday, D.D. Dominguez, J.A. Moran, W.D. Lee, and R. Eaton, *Synth. Met.* **9**, 397 (1984).

41   P.J. Barham and A. Keller, *J. Mater. Sci.*, **20**, 2281 (1985).

42   G. Capaccio, A.G. Gibson, and I.M. Ward, in *Ultra-High Modulus Polymers*, Applied Science Publishers, London, 1979, p. 1.

43   B. Kalb and A.J. Pennings, *J. Mater. Sci.*, **15**, 2584 (1980).

44   J. Smook and A.J. Pennings, *J. Mater. Sci.*, **19**, 31 (1984).

45   K.A. Hodd and D.C. Turley, *Chem. Br.* **14**, 545 (1978).

46   P.W. Morgan, *Plast. Rubber: Mater. Appl.*, **4**, 1 (Feb. 1979).

47   E.E. Magat, *Philos. Trans. R. Soc. London*, **A296**, 463 (1980).

48   S.L. Kwolek, P.W. Morgan, J.R. Schaefgen, and L.W. Gulrich, *Macromolecules*, **10**, 1390 (977).

49   D. Tanner, A.K. Dhingra, and J.J. Pigliacampi, *J. Met.*, **38**, 21 (Mar. 1986).

50   C.C. Chiao and T.T. Chiao, in *Handbook of Composites*, Van Nostrand Reinhold, New York, 1982, p. 272.

51   M. Jaffe and R.S. Jones, in *Handbook of Fiber Science & Technology*, vol. III, *High Technology Fibers*, Part A, Marcel Dekker, New York, 1985, p. 349.

52   M.G. Dobb, D.J. Johnson, and B.P. Saville, *Philos. Trans. R. Soc. London*, **A294**, 483 (1980).

53   A.R. West. *J. Mater. Sci.*, **16**, 2025 (1981).

54   S.J. DeTeresa, S.R. Allen, R.J. Farris, and R.S. Porter, *J. Mater. Sci.*, **19**, 57 (1984).

55   A.K. Dhingra, *Philos. Trans. R. Soc. London*, **A294**, 411 (1980).

56   K.S. Mazdiyasni, *Ceram. International*, **8**, 42 (1982).

57   H.E. DeBolt, V.J. Krukonis, and F.E. Wawner, in *Silicon Carbide — 1973*, University of South Carolina Press, Columbia, SC, 1974, p. 168.

58   S. Yajima, K. Okamura, J. Hayashi, and M. Omori, *J. Am. Ceram. Soc.*, **59**, 324 (1976)

59   S. Yajima, *Philos. Trans. R. Soc. London*, **A294**, 419 (1980).

60   K.J. Wynne and R.W. Rice, *Ann. Rev. Mater. Sci.*, **15**, 297 (1984).

61   C.-H. Andersson and R. Warren, *Composites*, **15**, 16 (Jan. 1984).

62   R. Warren and C.-H. Andersson, *Composites*, **15**, 101 (Apr. 1984).

63   S.G. Wax, *Am. Ceram. Soc. Bull.*, **64**, 1096 (1985).

64   K. Okamura, personal communication, 1986.
65   G. Simon and A.R. Bunsell, *J. Mater. Sci.*, **19**, 3649 (1984).
66   J.V. Milewski, J.L. Sandstrom, and W.S. Brown, in *Silicon Carbide — 1973*, University of South Carolina Press, Columbia, S C, 1974, p. 634.
67   J.-G. Lee and I. B. Cutler, *Am. Ceram. Soc. Bull.*, **54**, 195 (1975).
68   J.V. Milewski, F.D. Gac, J.J. Petrovic, S.R. Skaggs, *J. Mater. Sci.*, **20**, 1160 (1985).
69   J.J. Petrovic. J.V. Milewski. D.L. Rohr, and F.D. Gac, *J. Mater. Sci.*, **20**, 1167 (1985).
70   R.R. Wills, R.A. Mankle, and S.P. Mukherjee, *Am. Ceram. Soc. Bull.*, **62**, 904 (1983).
71   J. Economy and R. Lin, in *Boron and Refractory Borides*, Springer-Verlag, New York, 1977, p. 552.
72   A. Lindemanis, in *Emergent Process Methods for High Technology Ceramics*, Plenum Press, New York, 1983.
73   W.D. Smith, in *Boron and Refractory Borides*, Springer-Verlag, Berlin, 1977, p. 541.
74   R.A. Signorelli, in *Advances in Composite Materials*, Japan Society of Composite Materials, Tokyo, 1982, p. 37.
75   A. Kelly and H. Lilholt, *Philos. Mag.*, **20**, 311 (1969).
76   K.K. Chawla and M. Metzger, *J. Mater. Sci.*, **7**, 34 (1972).
77   K.K. Chawla, *Philos. Mag.*, **28**, 55 (1973).
78   K.K. Chawla and M. Metzger, *Met. Trans. A.*, **8A**, 1681 (1977).
79   D. Stöckel, in *Proceedings of the 1975 International Conference on Composite Materials*, TMS-AIME, New York, vol. 2, 1976, p. 484.
80   T.R. Anantharaman (ed.), *Metallic Glasses*, Trans. Tech. Pub., Aedermannsdorf, Switzerland, 1984, p. 1.
81   H.J. Guntherodt and H. Beck (eds.), *Metallic Glasses*, Springer-Verlag, Berlin, 1981.
82   C. Hargitai, I. Bakonyi, and T. Kemeny (eds.), *Metallic Glasses: Science & Technology*, Central Research Institute of Physics, Budapest, Hungary, 1981.
83   R. Hasegawa (ed.), *The Magnetic, Chemical, and Structural Properties of Glassy Metallic Alloys*, CRC Press, Boca Raton, FL, 1981.

Suggested Reading

P. Bracke, H. Schurmans, and J. Verhoest, *Inorganic Fibers and Composite Materials*, Pergamon Press, Oxford, 1983.

C.C. Chiao and T.T. Chiao, in *Handbook of Composites*, G. Lubin (ed.), Van Nostrand Reinhold. New York, 1982, p. 272.

T. Davis, H. Palmour, and T. Porter (eds.), *Emergent Process Methods for High Technology Ceramics*, Plenum Press, New York, 1982.

J. Delmonte, *Technology of Carbon and Graphite Fiber Composites*, Van Nostrand Reinhold, New York, 1981.

R.J. Diefendorf and E. Tokarsky, *Polym. Eng. Sci.*, **15**, 150 (1975).

J.B. Donnet and R.C. Bansal, *Carbon Fibers*, Marcel Dekker, New York, 1984.

E. Fitzer, *Carbon Fibres and Their Composites*, Springer-Verlag, Berlin, 1985.

M. Jaffe and R.S. Jones, High Performance Aramid Fibers, in *Handbook of Fiber Science and Technology*, vol. III, *High Technology Fibers*.

M. Langley (ed.), *Carbon Fibres in Engineering*, McGraw-Hill, London, 1973.

J. Preston, Aramid Fibers in *Kirk–Othmer Encyclopedia of Chemical Technology*, 3rd ed., vol. 4, Wiley-Interscience, New York, 1978.

W. Watt and B. V. Perov (eds.), *Strong Fibres*, vol 1. in the series Handbook of Composites, North-Holland, Amsterdam, 1985.

K.J. Wynne and R.W. Rice, Ceramics via Polymer Pyrolysis, *Ann. Rev. Mater. Sci.*, **14**, 297 (1984).

3. Matrix Materials

A brief description of the various matrix materials, polymers, metals, and ceramics, is given in this chapter. We emphasize the characteristics that are relevant to composites. The reader should consult the references listed under Suggested Reading for greater details regarding any particular aspect of these materials.

3.1 Polymers

Polymers are structurally much more complex than metals or ceramics. They are cheap and easily processible. On the other hand, polymers have lower strength and modulus and lower temperature use limits. Prolonged exposure to ultraviolet light and some solvents can cause the degradation of polymer properties. Because of predominantly covalent bonding, polymers are generally poor conductors of heat and electricity. Polymers, however, are generally more resistant to chemicals than are metals. Structurally, polymers are giant chainlike molecules (hence the name macromolecules is also used) with covalently bonded carbon atoms forming the backbone of the chain. The process of forming large molecules from small ones is called polymerization; that is, polymerization is the process of joining many monomers, the basic building blocks, together to form polymers. There are two important classes of polymerization:

1. Condensation polymerization: In this process there occurs a stepwise reaction of molecules and in each step a molecule of a simple compound, generally water, forms as a by-product.
2. Addition polymerization: In this process monomers join to form a polymer without producing any by-product. Addition polymerization is generally carried out in the presence of catalysts. The linear addition of ethylene molecules (CH_2) results in polyethylene (a chain of ethylene molecules), with the final mass of polymer being the sum of monomer masses:

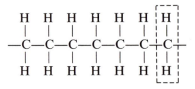

Two major classes of polymers produced by either condensation or addition polymerization, based on their behavior, are thermosetting and thermoplastic

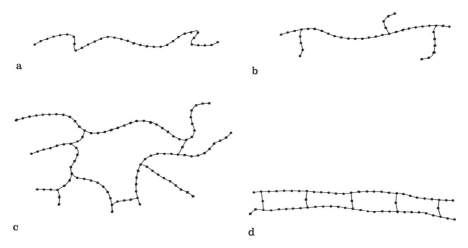

Fig. 3.1. Different molecular chain configurations: **a** linear, **b** branched, **c** crosslinked, **d** ladder

polymers. Their different behavior, however, stems from their molecular structure and shape, molecular size or mass, and the amount and type of bond (covalent or van der Waals). We first describe the basic molecular structure in terms of the configurations of chain molecules. Figure 3.1 shows the different chain configuration types.

1. *Linear Polymers.* As the name suggests, this type of polymer consists of a long chain of atoms with attached side groups. Examples include polyethylene, polyvinyl chloride, and polymethyl methacrylate. Figure 3.1a shows the configuration of linear polymers. Note the coiling and bending of the chain.
2. *Branched Polymers.* Polymer branching can occur with linear, crosslinked, or any other type of polymer; see Fig. 3.1b.
3. *Crosslinked Polymers.* In this case, molecules of one chain are bonded with those of another; see Fig. 3.1c. Crosslinking of molecular chains results in a three-dimensional network. Crosslinking makes sliding of molecules past one another difficult, thus making the polymers strong and rigid.
4. *Ladder Polymers.* If we have two linear polymers linked in a regular manner (Fig. 3.1d) we get a ladder polymer. Not unexpectedly, ladder polymers are more rigid than linear polymers.

Glass Transition Temperature

Pure, crystalline metals have well-defined melting temperatures. The melting point is the temperature at which crystalline order is completely destroyed on heating. Polymers, however, show a range of temperatures over which crystallinity vanishes. Figure 3.2 shows specific volume (volume/unit mass) versus temperature curves for amorphous and semicrystalline polymers. When a polymer liquid is cooled, it contracts. The contraction occurs because of a decrease in the thermal vibration of molecules and a reduction in the *free space*; that is, the molecules occupy the space less loosely. In the case of amorphous polymers, this contraction continues below

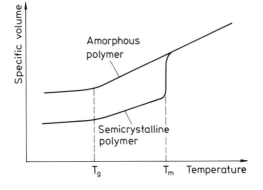

Fig. 3.2. Specific volume versus temperature for an amorphous and a semicrystalline polymer

T_m, the melting point of crystalline polymer, to T_g, the glass transition temperature where the supercooled liquid polymer becomes extremely rigid owing to extremely high viscosity. The structure below T_g is essentially that of a liquid. Since this is a phenomenon typical of glasses, we call it the glass transition temperature. It should be appreciated that in the case of amorphous polymers we are dealing with a glassy structure made of organic molecules. The glass transition temperature T_g is in many ways akin to the melting point for the crystalline solids. Many physical properties (e.g., viscosity, heat capacity, elastic modulus, and expansion coefficient) change abruptly at T_g. Polystyrene, for example, has a T_g of about 100°C and is therefore rigid at room temperature. Rubber, on the other hand, has a T_g of about — 75°C and therefore is flexible at room temperature. T_g is a function of the chemical structure of the polymer. For example, if a polymer has a rigid backbone structure and/or bulky branch groups, then T_g will be quite high.

Both amorphous polymers and glasses have a glass transition temperature T_g. The T_g of glasses, however, is several hundred degrees Celsius higher than that of polymers. The reason for this is the different types of bonding and the amount of crosslinking in the polymers and glasses. Glasses have mixed covalent and ionic bonding and are highly crosslinked. This gives them a higher thermal stability than polymers, which have covalent bonding only and a lesser amount of crosslinking than found in glasses.

Thermoplastics and Thermosets

Most linear polymers soften or melt on heating. These are called thermoplastic polymers and are suitable for liquid flow forming. Examples include low- and high-density polyethylene, polystyrene, and polymethyl methacrylate (PMMA). When the structure is amorphous, there is no apparent order among the molecules and the chains are arranged randomly; see Fig. 3.3a. Small, platelike single crystalline regions called lamellae or crystallites can be obtained by precipitation of the polymer from a dilute solution. In the lamellae long molecular chains are folded in a regular manner; see Fig. 3.3b. Many crystallites group together and form spherulites much like grains in metals.

When the molecules in a polymer are crosslinked in the form of a network, they do not soften on heating. We call these crosslinked polymers thermosetting poly-

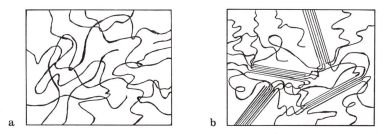

Fig. 3.3. Possible arrangements of polymer molecules: **a** amorphous, **b** semicrystalline

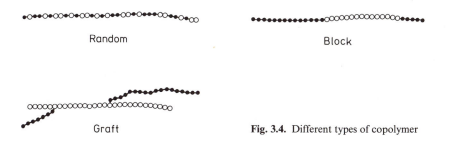

Random

Block

Graft

Fig. 3.4. Different types of copolymer

mers. Thermosetting polymers decompose on heating. Crosslinking makes sliding of molecules past one another difficult, thus making the polymer strong and rigid. A typical example is that of rubber crosslinked with sulfur, that is, vulcanized rubber. Vulcanized rubber has 10 times the strength of natural rubber. Common examples of thermosetting polymers include phenolic, polyester, polyurethane, and silicone.

Copolymers

There is another type of classification of polymers based on the type of repeating unit. When we have one type of repeating unit forming the polymer chain, we call it a homopolymer. Copolymers, on the other hand, are polymer chains having two different monomers. If the two different monomers are distributed randomly along the chain, then we have a regular or random copolymer. If, however, a long sequence of one monomer is followed by a long sequence of another monomer, we have a block copolymer. If we have a chain of one type of monomer and branches of another type, then we have a graft copolymer. Figure 3.4 shows schematically the different types of copolymer.

Molecular Weight

Molecular weight (MW) is a very important factor in polymers. Many mechanical properties increase with increasing molecular weight. In particular, resistance to deformation increases with molecular weight. Of course, concomitant with increasing molecular weight the processing of polymers becomes more difficult. The degree of polymerization (DP) indicates the number of basic units (mers) in a polymer. These two parameters are related as follows:

$$MW = DP \times (MW)_u$$

where $(MW)_u$ is the molecular weight of the repeating unit. Polymers do not have exactly identical molecules but have a mixture of different species, each of which has a different molecular weight or DP. Thus, the molecular weight of the polymer is characterized by a distribution function. Clearly, the narrower this distribution function is, the more homogeneous the polymer is. That is why one speaks of an average molecular weight or degree of polymerization.

Mandelkern [1] gives some interesting comparative values of low and high molecular weight materials which bring out the importance of molecular weight of polymers vis-à-vis monomeric materials. A molecule of water, H_2O, has a molecular weight of 18. Benzene, a low molecular weight organic solvent, has a molecular weight of 78. Compared to these, natural rubber has a molecular weight of about 10^6. Polyethylene, a common synthetic polymer, can have molecular weights greater than 10^5. The molecular size of these high molecular weight solids is also very large. The molecular diameter of water, for example, is 40 nm while that of polyethylene is about 6400 nm.

Degree of Crystallinity

Polymers can be amorphous or partially crystalline; see Fig. 3.3. A 100% crystalline polymer is an idealistic concept only. In practice, depending on the polymer type, molecular weight, and crystallization temperature, the amount of crystallinity in a polymer can vary from 30 to 90%. The inability to attain a fully crystalline structure is mainly due to the long chain structure of polymers. Some twisted and entangled segments of chains that get trapped between crystalline regions never undergo the conformational reorganization necessary to achieve a fully crystalline state. Molecular architecture also has an important bearing on the polymer crystallization behavior. Linear molecules with small or no side groups crystallize easily. Branched chain molecules with bulky side groups do not crystallize as easily. For example, linear high-density polyethylene can be crystallized to 90%, while branched polyethylene can be crystallized to only about 65%. Generally, the stiffness and strength of a polymer increase with the degree of crystallinity. It should be mentioned that deformation processes such as slip and twinning, as well as phase transformations that take place in monomeric crystalline solids, can occur in polymeric crystals as well.

Stress–Strain Behavior

Characteristic stress–strain curves of an amorphous polymer and of an elastomer (a rubbery polymer) are shown in Fig. 3.5a and b, respectively. Note that the elastomer does not show a Hookean behavior; its behavior is characterized as nonlinear elastic. The characteristically large elastic range shown by elastomers results from an easy reorganization of the tangled chains under the action of an applied stress.

Yet another point in which polymers differ from metals and ceramics is the extreme temperature dependence of their elastic moduli. Figure 3.6 shows schematically the variation of the elastic modulus of an amorphous polymer with temperature. In the temperature range below T_g, the polymer is hard and a typical

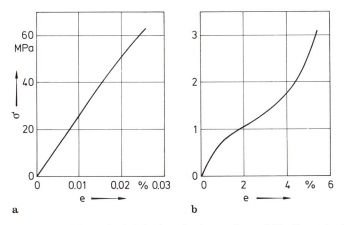

a   b

Fig. 3.5. a Hookean elastic behavior of a glassy polymer. **b** Nonlinear elastic behavior of an elastomer

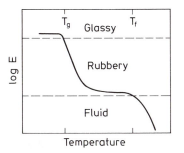

Fig. 3.6. Variation of elastic modulus of an amorphous polymer with temperature (schematic)

value of elastic modulus would be about 5 GPa. Above T_g, the modulus value drops significantly and the polymer shows a rubbery behavior. Above T_f (the temperature at which the polymer becomes fluid), the modulus drops abruptly. It is in this region of temperatures above T_f that polymers are subjected to various processing operations.

Thermal Expansion

Polymers generally have higher thermal expansivities than metals and ceramics. Furthermore, their thermal expansion coefficients are not truly constants; that is, the polymers expand markedly in a nonlinear way with temperature. Epoxy resins have coefficient of linear expansion values between 50×10^{-6} and 100×10^{-6} K^{-1} while polyesters show values between 100×10^{-6} and 200×10^{-6} K^{-1}. Small compositional changes can have a marked influence on the polymer expansion characteristics.

3.1.1 Common Polymeric Matrix Materials

Among the common polymer matrices used with continuous fibers are polyester and epoxy resins. The advantages of the former include their adequate resistance to water and a variety of chemicals, weathering, aging, and, last but not least, their very low cost. They can withstand temperatures up to about 80°C and they combine easily with glass fibers. Polyesters shrink between 4 and 8% on curing. A majority of common glass fiber reinforced composites has polyester as the matrix. A condensation reaction between a glycol (ethylene, propylene, diethylene glycol) and an unsaturated dibasic acid (maleic or fumaric) results in a linear polyester containing double bonds between certain carbon atoms. Crosslinking is provided by means of an addition reaction with styrene. Hardening and curing agents and ultraviolet absorbents are usually added. Thermosetting epoxy resins are more expensive than polyesters but they have better moisture resistance, lower shrinkage on curing (about 3%), a higher maximum use temperature, and good adhesion with glass fibers. A large number of proprietary formulations of epoxies is available and a very large fraction of high-performance polymer matrix composites has thermosetting epoxies as matrices. By far the most important variety of epoxies is a condensation product of epichlorohydrin and bisphenol A. Curing agents are organic amino or acid compounds, and crosslinking is obtained by introducing chemicals that react with the epoxy and hydroxy groups between adjacent chains. The extent of crosslinking is a function of the amount of curing agents. Generally, 10–15% by weight of amines or acid anhydrides is added and they become part of the epoxy structure. A detailed account of structure–property relationships of epoxies used as composite matrices is provided by Morgan [2].

Polyimides are thermosetting polymers that have a relatively high service temperature range, 250–300°C. But, like other thermosetting resins, they are brittle. Their fracture energies are in the $15–70$ J m^{-2} range. A major problem with polyimides is the elimination of water of condensation and solvents during the processing.

Bismaleimides (BMI), thermosetting polymers, can have service temperatures between 180 and 200°C. They have good resistance to hygrothermal effects. Being thermosets, they are quite brittle and must be cured at higher temperatures than conventional epoxies.

Thermoplastic resins are easier to fabricate than thermosetting resins; besides, they can be recycled. Heat and pressure are applied to form and shape them. More often than not, short fibers are used with thermoplastic resins but in the late 1970s continuous fiber reinforced thermoplastics began to be produced. The disadvantages of thermoplastics include their rather large expansion and contraction characteristics.

An important problem with polymer matrices is that associated with environmental effects. Polymers can degrade at moderately high temperatures and through moisture absorption. Absorption of moisture from the environment causes swelling in the polymer as well as a reduction in its T_g. In the presence of fibers bonded to the matrix, these hygrothermal effects can lead to severe internal stresses in the composite. The presence of thermal stresses resulting from thermal mismatch between matrix and fiber is of course a general problem in all kinds of composite materials. It is much more so in polymer matrix composites because polymers have high thermal expansivities.

Typical properties of some common polymeric matrix materials are summarized in Table 3.1 [3].

Table 3.1. Representative properties of some polymeric matrix materials

	Epoxy	Poly-imide	PEEK	Poly-amide-imide	Poly-ether-imide	Poly-sulfone	Poly-phenylene sulfide	Phenolics
Tensile strength (MPa)	35–85	120	92	95	105	75	70	50–55
Flexural modulus (MPa)	15–35	35	40	50	35	28	40	—
Density (g cm^{-3})	1.38	1.46	1.30	1.38	—	1.25	1.32	1.30
Continuous-service temperature (°C)	25–85	260–425	310	—	170	175–190	260	150–175
Coefficient of thermal expansion (10^{-5}°C^{-1})	8–11	9	—	6.3	5.6	9.4–10	9.9	4.5–11
Water absorption (24 h%)	0.1	0.3	0.1	0.3	0.25	0.2	0.2	0.1–0.2

Source: Adapted with permission from Ref. 3.

Matrix Toughness

Thermosetting resins (e.g., polyesters, epoxies, and polyimides) are highly cross-linked and provide adequate modulus, strength, and creep resistance but the same crosslinking of molecular chains causes extreme brittleness, that is, very low fracture toughness. By fracture toughness we mean resistance to crack propagation. It came to be realized in the 1970s that matrix fracture characteristics (strain to failure, work of fracture, or fracture toughness) are as important as lightness, stiffness, and strength properties. Figure 3.7 compares some common materials in terms of their fracture toughness as measured by the fracture energy in J/m^2 [4]. Note that thermosetting resins have values that are only slightly higher than those of glasses. Thermoplastic resins such as PMMA have fracture energies of about 1 kJ/m^2, while polysulfone thermoplastics have fracture energies of several 1 kJ/m^2, almost approaching those of the 7075-T6 aluminum alloy. Amorphous thermoplastic polymers show higher fracture energy values because they have a large free volume available which absorbs the energy associated with crack propagation. Among the well-known modified thermoplastics are the acrylonitrile–butadiene–styrene (ABS) copolymer and high-impact polystyrene (HIPS). One class of thermosetting resins that comes close to polysulfones is the elastomer-modified epoxies. Elastomer-modified or rubber-modified thermosetting epoxies form multiphase systems, a kind of composite in their own right. Small (a few micrometers or less), soft, rubbery inclusions distributed in a hard, brittle epoxy matrix enhance its toughness by several orders of magnitude [5–9].

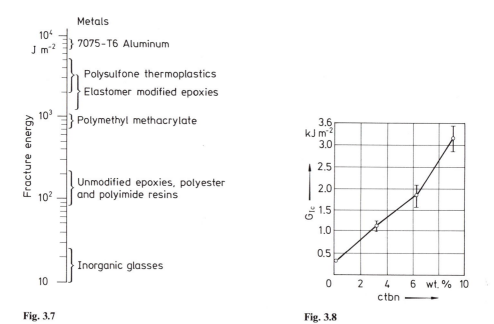

Fig. 3.7   Fig. 3.8

Fig. 3.7. Fracture energy for some common materials. (Adapted from Ref. 4, used with permission.)

Fig. 3.8. Fracture surface energy of an epoxy as a function of weight % of carboxyl-terminated butadiene–acrylonitrile (ctbn). (Adapted from Ref. 9, used with permission.)

Epoxy and polyester resins are commonly modified by introducing carboxyl-terminated butadiene–acrylonitrile copolymers (ctbn). Figure 3.8 shows the increase in fracture surface energy of an epoxy as a function of weight % of ctbn elastomer [9]. The methods of manufacture can be simple mechanical blending of the soft, rubbery particles and the resin or copolymerization of a mixture of the two.

Toughening of glassy polymers by elastomeric additions involves different mechanisms for different polymers. Among the mechanisms proposed for explaining this enhanced toughness are triaxial dilatation of rubber particles at the crack tip, particle elongation, and plastic flow of the epoxy. Ting [4] studied such a rubber-modified epoxy containing glass or carbon fibers. He observed that the mechanical properties of rubber-modified composite improved more in flexure than in tension. Scott and Phillips [9] obtained a large increase in matrix toughness by adding ctbn in unreinforced epoxy. But this large increase in toughness could be translated into only a modest increase in carbon fiber reinforced modified epoxy matrix composite. Introduction of a tough elastomeric phase, for example, a silicone rubber possessing good thermal resistance in a polyimide resin, produced a tough matrix material: a three- to five-fold gain in toughness G_{Ic} without a reduction in T_g [8].

Continuous fiber reinforced thermoplastics show superior toughness values owing to superior matrix toughness. Polyetheretherketone (PEEK) is a semicrystalline aromatic thermoplastic made by ICI in the United Kingdom [10–12] which is quite tough. Besides PEEK, there were developed in the 1980s some other tough thermoplastic resins, for example, K-polymers (Du Pont), which are a family of thermoplastic polyimides, and polyphenylene sulfide (PPS), which is a semicrystalline aromatic sulfide. Specifically, PEEK can have 20–40% crystalline phase. At 35% crystallinity, the spherulite size is about 2 μm [11]. Its glass transition temperature T_g is about 150°C and the crystalline phase melts at about 350°C. It has an elastic modulus of about 4 GPa, a yield stress of 100 MPa, and a fracture energy of about 500 J/m^2. The last item is particularly impressive. PPS is the simplest member of a family of polyarlene sulfides being developed by Phillips Petroleum [13]. PPS (trade name Ryton), a semicrystalline polymer, has been reinforced by chopped carbon fibers as well as prepregged with continuous carbon fibers at Phillips Petroleum [13].

3.2 Metals

Metals are, by far, the most versatile of engineering materials. They are strong and tough. They can be plastically deformed and can be strengthened by a wide variety of methods involving mostly obstruction of movement of linear defects called dislocations. For a detailed treatment of the structure and properties of metals, the reader is referred to selected publications in this area listed under the Suggested Reading section at the end of the chapter. We present here a summary highlighting the salient features of metals and a brief analysis as to why metals should be reinforced with high-modulus but low-density fibers.

3.2.1 Structure

Metals, with the exception of metallic glasses, are crystalline materials. The vast majority of metals exists in one of the following three crystalline forms:

1. Face-centered cubic (fcc)
2. Body-centered cubic (bcc)
3. Hexagonal close packed (hcp)

Figure 3.9 shows these three structures. The black dots mark the centers of the atomic positions while some of the atomic planes are shaded. The atoms in fact touch each other and all space is filled up. Some important metals with their respective crystalline structures are listed in Table 3.2. Metals are crystalline materials; however, the crystalline structure is never perfect. Metals contain a variety of crystal imperfections. We can classify these as follows:

1. Point defects (zero dimensional)
2. Line defects (unidimensional)
3. Planar or interfacial defects (bidimensional)
4. Volume defects (tridimensional)

Point defects can be of three types. A vacancy is created when an atomic position in the crystal lattice is vacant. An interstitial is produced when an atom of the

Plane (111)   Plane (110)   Plane (0001)

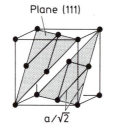 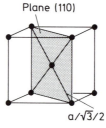 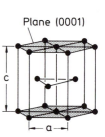

$a/\sqrt{2}$ $a/\sqrt{3}/2$ $\leftarrow a \rightarrow$

a Face-centered cubic   **b** Body-centered cubic   **c** Hexagonal close packed

Fig. 3.9a–c. Three crystalline forms of metals

Table 3.2. Crystal structure of some important metals

fcc	bcc	hcp
Iron (910–1390°C)	Iron ($T < 910°C$ and $T > 1390°C$)	Titanium[a]
Nickel	Beryllium ($T > 1250°C$)	Beryllium ($T < 1250°C$)
Copper	Cobalt ($T > 427°C$)	Cobalt ($T < 427°C$)
Aluminum	Tungsten	Cerium ($-150°C < T < -10°C$)
Gold	Molybdenum	Zinc
Lead	Chromium	Magnesium
Platinum	Vanadium	Zirconium[a]
Silver	Niobium	Hafnium[a] ($T < 1950°C$)

[a] Undergoes bcc ⇌ hcp transformation at different temperatures.

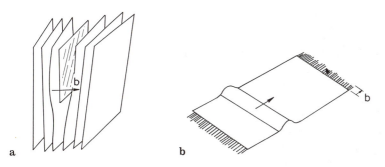

Fig. 3.10. a Edge dislocation. b Dislocation in a carpet

material or a foreign atom occupies an interstitial or nonlattice position. A substitutional point defect comes into being when a regular atomic position is occupied by a foreign atom. Intrinsic point defects (vacancies and self-interstitials) in metals exist at a given temperature in equilibrium concentrations. Increased concentrations of these defects can be produced by quenching from high temperatures, bombarding with energetic particles such as neutrons, and plastic deformation. Point defects can have a marked effect on the mechanical properties.

Line imperfections, called dislocations, have been one of the critically important structural imperfections in the area of physical and mechanical metallurgy, diffusion, and corrosion. A dislocation is defined by two vectors: a dislocation line vector (tangent to the line) and its Burgers vector, which gives the direction of atomic displacement. The dislocation has a kind of lever effect because it allows by its movement one part of metal to be sheared over the other without the need for simultaneous movement of atoms across a plane. It is the presence of these line imperfections that makes it easy to deform metals plastically. Under normal circumstances then, the plastic deformation of metals is accomplished by the movement of these dislocations. Figure 3.10a shows an edge dislocation. The two vectors, defining the dislocation line and the Burgers vector, are designated as t (not shown in Fig. 3.10a) and b, respectively. The dislocation permits shear in metals at stresses much below those required for simultaneous shear across a plane. Figure 3.10b shows this in an analogy. A carpet of course can be moved by pushing or pulling. However, a much lower force is required to move the carpet by a distance b if a defect is introduced into it and made to move the whole extension of the carpet. Figure 3.11 shows dislocations as seen by transmission electron microscopy in a thin foil of steel. This is a dark field electron micrograph and dislocation lines appear as white lines. Dislocations become visible in the transmission electron microscope because of the distortion of the atomic planes owing to the presence of dislocations. Also to be seen in this micrograph are equiaxial precipitate particles pinning the dislocations at various points.

The interfacial or planar defects occupy an area or surface of the crystal, for example, grain boundaries, twin boundaries, domains, or antiphase boundaries. Grain boundaries are, by far, the most important of these planar defects from the mechanical metallurgy point of view. Among the volumetric or tridimensional defects we can include large inclusions and gas porosity existing cracks.

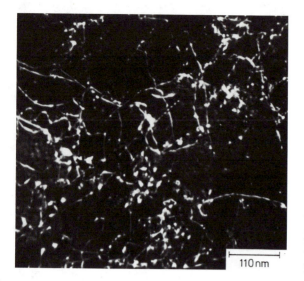

Fig. 3.11. Dislocations (*white lines*) in a steel sample: dark field transmission electron micrograph. Also visible are equiaxial precipitate particles

110 nm

3.2.2 Conventional Strengthening Methods

Experimental results show that work hardening (or strain hardening), which is the ability of a metal to become more resistant to deformation as it is deformed, is related in a singular way to the dislocation density (ρ) after deformation. There exists a linear relationship between the flow stress τ and $\sqrt{\rho}$ [14]:

$$\tau = \tau_0 + \alpha GB \sqrt{\rho} \tag{3.1}$$

where τ_0 is the shear stress required to move a dislocation in the absence of any other dislocations. Basically, work hardening results from the interactions among dislocations moving on different slip planes. A tangled dislocation network results after a small plastic deformation, which impedes the motion of other dislocations. This, in turn, requires higher loads for further plastic deformation. Various theories [15, 16] explain the interactions of dislocations with different kinds of barrier (e.g., dislocations, grain boundaries, solute atoms, and precipitates) that result in characteristic strain hardening of metals. All these theories arrive at the relationship between τ and ρ given in Eq. (3.1), indicating that a particular dislocation distribution is not crucial and that strain hardening in practice is a statistical result of some factor that remains the same for various distributions. Cold working of metals, which leads to the strengthening of metals as a result of work hardening, is a routinely used strengthening technique.

A similar relationship exists between the flow stress τ and the mean grain size (or dislocation cell size) to an undetermined level

$$\tau = \tau_0 + \frac{\alpha' Gb}{D^{1/2}} \tag{3.2}$$

where D is the mean grain diameter. This relationship is known as the Hall–Petch relationship after the two researchers who first postulated it [17, 18]. Again, various models have been proposed to explain this square root dependence on grain size. Earlier explanations involved a dislocation pile-up bursting through the boundary owing to stress concentrations at the pile-up tip and activation of dislocation sources in adjacent grains [19]. Later theories involved the activation of grain boundary dislocations into grain interiors, elastic incompatibility stresses between adjacent grains leading to localized plastic flow at the grain boundaries, and so on [20, 21]. An important aspect of strengthening by grain refinement is that, unlike other strengthening mechanisms, it results in an improvement in toughness concurrent with that in strength (again to an undefined lower grain size). Another easy way of strengthening metals by impeding dislocation motion is that of introducing heterogeneities such as solute atoms or precipitates or hard particles in a ductile matrix. When we introduce solute atoms (say, carbon, nitrogen, or manganese in iron) we obtain solid solution hardening. Interstitial solutes such as carbon and nitrogen are much more efficient strengthening agents than substitutional solutes such as manganese and silicon. This is so because the interstitials cause a tetragonal distortion in the lattice and thus interact with both screw and edge dislocations, while the substitutional atoms cause a spherical distortion that interacts only with edge dislocations, because the screw dislocations have a pure shear stress field and no hydrostatic component. Precipitation hardening of a metal is obtained by decomposing a supersaturated solid solution to form a finely distributed second phase. Classical examples of precipitation strengthening are those of Al-Cu and Al-Zn-Mg alloys which are used in the aircraft industry [22]. Oxide dispersion strengthening involves artificially dispersing rather small volume fractions (0.5–3 vol.%) of strong and inert oxide particles (e.g., Al_2O_3, Y_2O_3, and ThO_2) in a ductile matrix by internal oxidation or powder metallurgy blending techniques [23]. Both the second-phase precipitates and dispersoids act as barriers to dislocation motion in the ductile matrix, thus making the matrix more deformation resistant. Dispersion-hardened systems (e.g., $Al + Al_2O_3$) show high strength levels at elevated temperatures while precipitates (say, $CuAl_2$ in aluminum) tend to dissolve at those temperatures. Precipitation hardening systems, however, have the advantage of enabling one to process the alloy in a soft condition and giving the precipitation treatment to the finished part. The precipitation process carried out for long periods of time can also lead to overaging and solution, that is, a weakening effect.

Quenching a steel to produce a martensitic phase has been a time honored strengthening mechanism for steels. The strength of the martensite phase in steel depends on a variety of factors, the most important being the amount of carbon. The chemical composition of martensite is the same as that of the parent austenite phase from which it formed, but it is supersaturated with carbon [24]. Carbon saturation and the lattice distortion that accompanies the transformation lead to the high hardness and strength of martensite.

A novel approach to obtaining enhanced mechanical performance is rapid solidification processing [25]. By cooling metals at rates in the 10^4–10^9 K s^{-1} range, it is possible to produce microstructures that are unique. Very fine powders or ribbons of rapidly solidified materials are processed into bulk materials by hot pressing, hot isostatic pressing, or hot extrusion. The rapidly solidified materials can be amorphous (noncrystalline), microcrystalline (grain size below 1 μm), microdendritic solid solutions containing solute concentrations vastly superior to those of conventionally processed materials. Effectively, massive second-phase particles are totally eliminated. These unique microstructures lead to very favorable mechanical properties.

3.2.3 Properties of Metals

Typical values of elastic modulus, yield strength, and ultimate strength in tension, as well as those of fracture toughness of some common metals and their alloys, are listed in Table 3.3, while typical engineering stress–strain curves in tension are shown in Fig. 3.12. Note the rather large plastic strain range.

3.2.4 Why Fiber Reinforcement of Metals?

Precipitation or dispersion hardening of a metal can result in a dramatic increase in the yield stress and/or the work hardening rate. The influence of these obstacles on the elastic modulus is negligible. This is so because the intrinsic properties of the strong particles (e.g., the high elastic modulus) are not utilized. Their only function

Table 3.3. Mechanical properties of some common metals and alloys

	E (GPa)	σ_y (MPa)	σ_{max} (MPa)	K_{Ic} (MPa m$^{1/2}$)
Pure (ductile) metals				
Aluminum	70	40	200	100
Copper	120	60	400	to
Nickel	210	70	400	350
Ti-6A1-4V	110	900	1000	120
Aluminum alloys (high strength–				
low strength)	70	100–380	250–480	23–40
Plain carbon steel	210	250	420	140
Stainless steel (304)	195	240	365	200

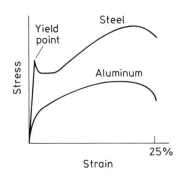

Fig. 3.12. Stress–strain curves of two common metals. Note the large plastic strain range.

Table 3.4. Density of some common metals

Metal	Density (g cm^{-3})	Metal	Density (g cm^{-3})
Aluminum	2.7	Lead	11.0
Beryllium	1.8	Nickel	8.9
Copper	8.9	Silver	10.5
Gold	19.3	Titanium	4.5
Iron	7.9	Tungsten	19.3

is to impede dislocation movements in the metal. The improvement in stiffness can be profitably obtained by incorporating so-called advanced high-modulus fibers in a metal matrix. It turns out that most of these high-modulus fibers are also lighter than the metallic matrix materials; the only exception is tungsten which has a high modulus and is also very heavy. Table 3.4 lists some common metals and their densities ρ. The densities of various fibers are given in Chap. 2.

Although one generally thinks of having a high Young's modulus as something very desirable from a structural point of view, it turns out that for structural applications involving compression or flexural loading, for example, of beams (say, in a plane, rocket, or truck), it is the E/ρ^2 value that should be maximized. Consider a simple square section cantilever beam, of length l, thickness t, and under an applied force P. The elastic deflection of this beam, ignoring self-weight, is given by [26]

$$\delta = \frac{Pl^3}{3\,EI}$$

where I is the moment of inertia; in this case it is equal to $t^4/12$.

Therefore,

$$\delta = \frac{4\,l^3 P}{Et^4} \tag{3.3}$$

The mass of this beam is

$$M = \text{volume} \times \text{density} = lt^2\rho$$

or

$$t = \left(\frac{M}{l\rho}\right)^{1/2} \tag{3.4}$$

From Eqs. (3.3) and (3.4) we have

$$\delta = \left(\frac{4\,l^3 P}{E}\right)\left(\frac{l^2\rho^2}{M^2}\right)$$

or

$$M = \left(\frac{4\,l^5 P}{\delta}\right)\left(\frac{\rho^2}{E}\right)^{1/2}$$

Thus, for a given rigidity or stiffness P/δ, we have a minimum of mass when the parameter E/ρ^2 is a maximum. What this simple analysis shows is that it makes good sense to use high-modulus fibers to reinforce metals in a structural application and it makes eminently more sense to use fibers that are not only stiffer than metallic matrices but also lighter.

3.3 Ceramic Matrix Materials

Ceramic materials are very hard and brittle. They consist of one or more metals combined with a nonmetal such as oxygen, carbon, nitrogen, or boron. They have strong ionic bonds and very few slip systems available compared to metals. These result in characteristically low failure strains and low toughness or fracture energies. Besides being brittle, they lack uniformity in properties, have low thermal and mechanical shock resistance, and have low tensile strength. On the other hand, ceramic materials have very high elastic moduli, low densities, and can withstand very high temperatures. The last item is very important and is the real driving force behind the effort to produce tough ceramics. Consider the fact that metallic super-alloys, used in jet engines, can easily withstand temperatures up to 800°C and can go up to 1000°C with oxidation-resistant coatings. Beyond this temperature, one must resort to ceramic materials.

But, by far, the major disadvantage of ceramics is their extreme brittleness. Even the minutest of surface flaws (scratches or nicks) or internal flaws (inclusions, pores, or microcracks) can have disastrous results. Understandably then, a good deal of research effort in the 1970s and 1980s has gone toward making tougher ceramics. One important approach in this regard is that of fiber reinforcement of brittle ceramics. We shall describe the ceramic matrix composites in Chapter 7. Here we make a brief survey of ceramic materials, emphasizing the ones that are commonly used as matrices.

3.3.1 Bonding and Structure

Ceramic materials, with the exception of glasses, are crystalline, as are metals. Unlike metals, however, they have mostly ionic bonding and some covalent bonding. Ionic bonding involves electron transfer between atomic species constituting the ceramic compound; that is, one atom gives up an electron(s) while another accepts an electron(s). Electrical neutrality is maintained; that is, positively charged ions (cations) balance the negatively charged ions (anions). Generally, ceramic compounds are stoichiometic; that is, there exists a fixed ratio of cations to anions. Examples are alumina (Al_2O_3), beryllia (BeO), spinels ($MgAl_2O_4$), silicon carbide (SiC), and silicon nitride (Si_3N_4). It is not uncommon, however, to have nonstoichiometric ceramic compounds, for example, $Fe_{0.96}O$. The oxygen ion (anion) is generally very large compared to the metal ion (cation). Thus, cations occupy interstitial positions in a crystalline array of anions.

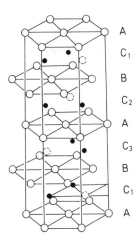

○ Oxygen
● Aluminum
◌ Vacant

Fig. 3.13. Hexagonal closed-packed structure of α-alumina. A and B layers contain oxygen atoms. C_1, C_2, and C_3 contain aluminum atoms. The C layers are only two-thirds full

Crystalline ceramics generally exhibit simple cubic, close-packed cubic, and hexagonal close-packed structures. Simple cubic structure is also called the cesium chloride structure. It is, however, not very common. CsCl, CsBr, and CsI show this structure. The two species form an interpenetrating cubic array with anions occupying the cube corner positions while cations go to the interstitial sites. Cubic close packed is really a variation of the fcc structure described in Sect. 3.2. Oxygen ions (anions) make the proper fcc structure with metal ions (cations) in the interstices. Many ceramic materials show this structure, also called the NaCl or rock salt type structure. Examples include MgO, CaO, FeO, NiO, MnO, and BaO. There are other variations of fcc close-packed structures, for example, zinc blende types (ZnS) and fluorite types (CaF). The hexagonal close-packed structure is also observed in ceramics. ZnS, for example, also crystallizes in the hcp form. Other examples are nickel arsenide (NiAs) and corundum (Al_2O_3). Figure 3.13 shows the hcp crystal structure of α-Al_2O_3. A and B layers consist of oxygen atoms while C_1, C_2, and C_3 layers contain aluminum atoms. The C layers are only two-thirds full.

Glass-ceramic materials form yet another important category of ceramics. They form a sort of composite material because they consist of 95–98% by volume of crystalline phase and the rest is glassy phase. The crystalline phase is very fine (grain size less than 1 μm diameter). Such a fine grain size is obtained by adding nucleating agents (commonly TiO_2 and ZrO_2) during the melting operation followed by controlled crystallization. Important examples of glass-ceramic systems include:

1. Li_2O-Al_2O_3-SiO_2. This has a very low thermal expansion and is therefore very resistant to thermal shock. Corning ware is a well-known trade name of this class of glass-ceramic.
2. MgO-Al_2O_3-SiO_2. This has high electrical resistance coupled with high mechanical strength.

Ceramic materials can also form solid solutions. Unlike metals, however, interstitial solid solutions are less likely in ceramics because the normal interstitial sites are already filled. Introduction of solute ions disrupts the charge neutrality. Vacancies accommodate the unbalanced charge. For example, FeO has a NaCl type structure with an equal number of Fe^{2+} and O^{2-} ions. If, however, two Fe^{3+} ions were to replace three Fe^{2+} ions, we would have a vacancy where an iron ion would form.

Glasses, the traditional silicate ceramic materials, are inorganic solidlike materials that do not crystallize when cooled from the liquid state. Their structure (see Fig. 2.5) is not crystalline but that of a supercooled liquid. In this case we have a specific volume versus temperature curve similar to the one for polymers (Fig. 3.2) and a characteristic glass transition temperature T_g. Under certain conditions, crystallization of glass can occur with an accompanying abrupt decrease in volume at the melting point because the atoms take up ordered positions.

3.3.2 Effect of Flaws on Strength

Just as in metals, imperfections in crystal packing of ceramics do exist and reduce their strength. The difference is that important defects in ceramic materials are surface flaws and vacancies. Dislocations do exist but are relatively immobile. Grain boundaries and free surfaces are important planar defects. As in metals, small grain size improves the mechanical properties of ceramics.

Surface flaws and internal pores (Griffith flaws) are particularly dangerous for strength and fracture toughness of ceramics. Fracture stress for an elastic material having an internal crack of length $2a$ is given by the Griffith relationship, namely,

$$\sigma_f = \left(\frac{2E\gamma}{a}\right)^{1/2}$$

where E is the Young's modulus and γ is the surface energy of the crack surface. Linear elastic fracture mechanics treats this problem of brittle fracture in terms of a parameter called the stress intensity factor, K. The stresses at the crack tip are given by

$$\sigma_{ij} = \left(\frac{K}{(2\pi r)^{1/2}}\right)f_{ij}(\theta)$$

where $f_{ij}(\theta)$ is a function of the angle θ. Fracture occurs when K attains a critical value K_{Ic}. Yet another approach is based on the energy viewpoint, a modification of the Griffith idea. Fracture occurs, according to this approach, when the crack extension force G reaches a critical value G_{Ic}. For ceramic materials, $G_{Ic} = 2\gamma$. It can also be shown that $K^2 = EG$, for opening failure mode and plane stress; that is, the stress intensity factor and the energy approaches are equivalent.

3.3.3 Common Ceramic Matrix Materials

Silicon carbide has excellent high-temperature resistance. The major problem is that it is quite brittle up to very high temperatures and in all environments. Titanium diboride and silicon nitride are the other important nonoxide ceramic matrix materials. Among the oxide ceramics, alumina and zirconia are quite promising. Glass, oxy-nitride glasses, and glass-ceramics are other ceramic matrices. With

Table 3.5. Properties of some ceramic matrix materials

	Young's modulus (GPa)	Tensile strength (MPa)	Coefficient of thermal expansion (10^{-6} K^{-1})	Density (g cm^{-3})
Borosilicate glass	60	100	3.5	2.3
Soda glass	60	100	8.9	2.5
Lithium aluminosilicate glass-ceramic	100	100–150	1.5	2.0
Magnesium aluminosilicate glass-ceramic	120	110–170	2.5–5.5	2.6–2.8
Mullite	143	83	5.3	
MgO	210–300	97–130	13.8	3.6
Si$_3$N$_4$	310	410	2.25–2.87	3.2
Al$_2$O$_3$	360–400	250–300	8.5	3.9–4.0
SiC	400–440	310	4.8	3.2

Source: Adapted with permission from Ref. 27.

glass-ceramics one can densify the matrix in a glassy state with fibers, followed by crystallization of the matrix to obtain high-temperature stability.

Ceramic matrices are employed in fiber reinforced composites with the objective of achieving, besides high strength and stiffness, high-temperature stability and adequate fracture toughness. Table 3.5 summarizes some of the important characteristics of common ceramic matrix materials [27].

References

1 L. Mandelkern, *An Introduction to Macromolecules*, 2nd ed., Springer-Verlag, New York, 1983, p. 1.
2 R.J. Morgan, in *Epoxy Resins and Composites I*, Springer-Verlag, Berlin, 1985, p. 1.
3 L.K. English, *Mater. Eng.*, **102**, 32 (Sept. 1985).
4 R.Y. Ting, in *The Role of Polymeric Matrix in the Processing and Structural Properties of Composite Materials*, Plenum Press, New York, 1983, p. 171.
5 J.N. Sultan and F.J. McGarry, *Polym. Eng. Sci.*, **13**, 29 (1973).
6 C.K. Riew, E.H. Rowe, and A.R. Siebert, in *Toughness and Brittleness of Plastics*, American, Chemical, Society Advances in Chemistry series, vol. 154 1976, p. 326.
7 W.D. Bascom and R.L. Cottington, *J. Adhesion*, **7**, 333 (1976).
8 A.K. St. Clair and T.L. St. Clair, *Int. J. Adhesion and Adhesives*, **1**, 249 (1981).
9 J.M. Scott and D.C. Phillips, *J. Mater. Sci.*, **10**, 551 (1975).
10 J.T. Hartness, *Sampe Q.*, **14**, 33 (Jan. 1983).
11 F.N. Cogswell, *Sampe Q.*, **14**, 33 (July 1983).
12 D.J. Blundell, J.M. Chalmers, M.W. Mackenzie, and W.F. Gaskin, *Sampe Q.*, **16**, 22 (July 1985).
13 J.E. O'Connor, W.H. Beever, and J.F. Geibel, *Proc. Sampe Int. Symp.*, **31**, 1313 (1986).
14 H. Wiedersich, *J. Met.*, **10**, 425 (1964).
15 A. Seeger, in *Dislocations and Mechanical Properties of Crystals*, John Wiley & Sons, New York, 1957, p. 23.
16 D. Kuhlmann-Wilsdorf, in *Work Hardening in Tension and Fatigue*, TMS-AIME, New York, 1977, p. 1.

17   E.O. Hall, *Proc. R. Soc. London*, **B64**, 474 (1951).

18   N.J. Petch, *J. Iron Steel Inst.*, **174**, 25 (1953).

19   A.H. Cottrell, *Trans. TMS-AIME*, **212**, 192 (1958).

20   J.C.M. Li, *Trans. TMS-AIME*, **227**, 239 (1963).

21   M.A. Meyers and E. Ashworth, *Philos. Mag.*, **46**, 737 (1982).

22   M.E. Fine, *Phase Transformations in Condensed Systems*, Macmillan, New York, 1964.

23   G.S. Ansell (ed.), *Oxide Dispersion Strengthening*, Gordon & Breach, New York, 1968.

24   M.J. Roberts and W.S. Owen, *J. Iron Steel Inst.*, **206**, 375 (1968).

25   N.J. Grant, in *Frontiers in Materials Technologies*, Elsevier, New York, 1985, p. 125.

26   R.W. Fitzgerald, *Mechanics of Materials*, 2nd ed., Addison-Wesley, Reading, MA, 1982, p. 205.

27   D.C. Phillips, in *Fabrication of Composites*, North-Holland, Amsterdam, 1983, p. 373.

Suggested Reading

C.S. Barrett and T.B. Massalski, *Structure of Metals*, 3rd ed. Pergamon Press, Oxford, 1980.

K. Dusek (ed.), *Epoxy Resins and Composites I* (Advances in Polymer Science, vol. 72), Springer-Verlag, Berlin, 1985.

A.G. Evans and T.G. Langdon, Structural Ceramics, *Prog. Mater. Sci.*, **21**, 171 (1976).

W.D. Kingery, H.K. Bowen, and D.R. Uhlmann, *Introduction to Ceramics*, 2nd ed., John Wiley & Sons, New York, 1976.

I. LeMay, *Principles of Mechanical Metallurgy*, Elsevier, New York, 1981.

L. Mandelkern, *An Introduction to Macromolecules*, 2nd ed., Springer-Verlag, New York, 1983.

M.A. Meyers and K.K. Chawla, *Mechanical Metallurgy: Principles and Applications*, Prentice-Hall, Englewood Cliffs, N.J., 1984.

R.F. Reed-Hill, *Physical Metallurgy Principles*, 2nd ed., Van Nostrand, 1973.

D.W. Richerson, *Modern Ceramic Engineering*, Marcel Dekker, New York, 1982.

J.C. Seferis and L. Nicolais (eds.), *The Role of Polymeric Matrix in the Processing and Structural Properties of Composite Materials*, Plenum Press, New York, 1983.

4. Interfaces

We can define an interface between any two phases, say fiber and matrix, as a bounding surface where a discontinuity of some kind occurs. The discontinuity may be sharp or gradual. In general, the interface is an essentially bidimensional region through which material parameters, such as concentration of an element, crystal structure, atomic registry, elastic modulus, density, and coefficient of thermal expansion, change from one side to another. Clearly, a given interface may involve one or more of these items. Most of the physical, chemical, or mechanical discontinuities listed are self-explanatory. The concept of atomic registry perhaps needs some further elaboration. In terms of the atomic registry types, we can have a coherent, semicoherent, or incoherent interface. A coherent interfaces is one where atoms at the interface form part of both the crystal lattices; that is, there exists a one-to-one correspondence between atomic sites on the two sides of the interface. In general, a perfect atomic registry does not occur between unconstrained crystals. Rather, coherency at the interface invariably involves an elastic deformation of the crystals. A coherent interface, however, has a lower energy than an incoherent one. A classic example of coherent interface is the interface between G-P zones and the aluminum matrix. With increasing size of the crystals, the elastic strain energy becomes more than the interfacial energy, leading to a lowering of the free energy of the system by introduction of dislocations at the interface. Such an interface, containing dislocations to accommodate the large interfacial strains and thus having only a partial atomic registry, is called a semicoherent interface. As examples, we cite interfaces between a precipitate and a matrix as well as some eutectic composites; for example, NiAl-Cr system [1] has semicoherent interfaces between phases. With still further increases in crystal sizes, the dislocation density at the interface increases, and eventually the dislocations lose their distinct identity; that is, it is no longer possible to specify individual atomic positions at the interface. Such an interface is called an incoherent interface. In general, such a boundary has a transition zone, a few atomic diameters wide, in which the atomic positions cannot be specified. The atoms located at such an interface do not correspond to the structure of either one of the two crystals or grains joining at the grain boundary.

More often than not, the interface between fiber and matrix is rather rough instead of the ideal planar interface; see Fig. 4.1. Under such a situation, it is useful to discuss the intimate contact between the fiber and matrix in terms of the wettability concept. Particularly in the case of the polymer matrix composites, an intimate contact at the molecular level between the fiber and the matrix brings intermolecular forces into play with or without causing a chemical linkage between the components. This intimate contact between the fiber and the matrix requires

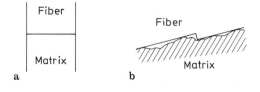

Fiber

Matrix

a

Fiber

Matrix

b

Fig. 4.1. **a** An ideal planar interface between fiber and matrix. **b** A more likely real, jagged interface between fiber and matrix

that the latter in liquid form must wet the former. Coupling agents are frequently used to improve the wettability between the components. At times, other approaches, such as modifying the matrix composition, are employed. In what follows we^c describe the various aspects of interfaces in composites in general. Characteristics of interfaces in specific composite systems are discussed later in Chaps. 5–9.

4.1 Wettability and Bonding

Baier et al. [2] have discussed the mechanisms that assist or impede adhesion. A key concept in this regard is that of wettability. Wettability is the term used to describe the extent to which a liquid would spread on a solid surface. We can measure the wettability of a given solid by a liquid by considering the equilibrium of forces in a system consisting of a drop resting on a plane solid surface in the right atmosphere. The liquid drop will spread and wet the surface completely only if this results in a net reduction of the system free energy. Figure 4.2 shows the situation schematically. Note that a portion of the solid–vapor interface is substituted by the solid–liquid interface. Then, we can write the condition for complete wettability as

$$\gamma_{LS} + \gamma_{LV} < \gamma_{SV} \tag{4.1}$$

where γ represents the specific surface energy while the subscripts S, L, and V denote solid, liquid, and vapor, respectively. If this process of substitution of the solid–vapor interface involves an increase in the free energy of the system, then complete spontaneous wetting will not result. Under such conditions, the liquid will spread until a balance of forces acting on the surface is attained; that is, we shall have partial wetting. The angle θ that the liquid drop makes with the solid surface is called the contact angle (Fig. 4.2). From the equilibrium of forces, we can write

$$\gamma_{LS} + \gamma_{LV} \cos \theta = \gamma_{SV} \tag{4.2}$$

or

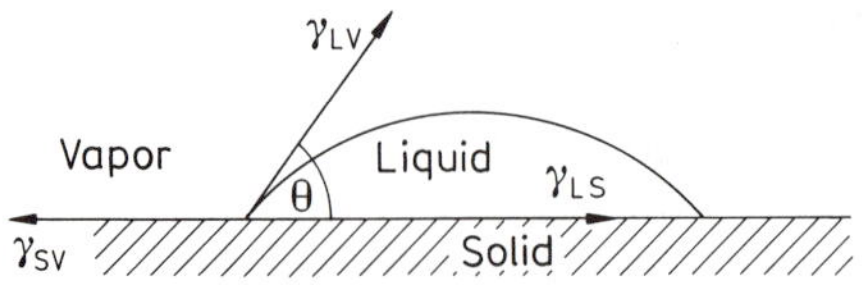

γ_{LV}

Vapor

Liquid

γ_{LS}

θ

γ_{SV} Solid

Fig. 4.2. A liquid drop on a solid surface making a contact angle between the solid and the liquid. The terms γ_{LS}, γ_{LV}, and γ_{SV} denote the surface energies of solid–liquid, liquid–vapor, and solid–vapor interfaces, respectively

$$\theta = \cos^{-1}\left(\frac{\gamma_{SV} - \gamma_{LS}}{\gamma_{LV}}\right) \tag{4.3}$$

For an angle of 180°, the drop assumes a spherical shape and no wetting occurs. Conversely, a 0° contact angle represents perfect wetting. For $0° < \theta < 90°$, a partial wetting occurs. Not unexpectedly, the contact angle depends on the nature of the surfaces, whether or not absorbed gases or oxide films are present, and so on. Surface roughness would diminish the contact angle while absorbed gases would increase it. Any impurities in or deliberate additions to the solid or liquid phase or a chemical reaction between the phases would affect the wettability.

The terms bonding and wettability have been used rather loosely in the literature. Good bonding implies that atomic or molecular bonds are formed uniformly all along the interface. The bond strength, however, may vary from weak van der Waals to strong covalent bonds. An intimate contact at the atomic or molecular level aids in bonding. Wettability refers to the extent of intimate contact possible at the molecular level. As mentioned above, the term is used to describe the extent to which a liquid will wet a solid. A low contact angle indicates good wettability, while a high contact angle indicates poor wettability.

4.2 The Interface in Composites

The behavior of a composite material is a result of the combined behavior of the following:

1. Fiber or the reinforcing element
2. Matrix
3. The interface between the fiber and the matrix

To obtain desirable characteristics in the composite material, one should be careful that the fibers are not weakened by flaws (surface or internal) and that the applied load is effectively transferred from the matrix to the fibers via the interface. Clearly, the interface generally has an important bearing in this regard. Specifically, in the case of a fiber reinforced composite material, the interface, or more precisely the interfacial zone, consists of near surface layers of fiber and matrix and any layer(s) of material existing between these surfaces. The reason the interface in a composite is of great importance is that the internal surface area occupied by the interface is quite extensive. It can easily go as high as 3000 cm^2/cm^3 in a composite containing a reasonable fiber volume fraction. Thus, it becomes extremely important to understand what exactly is going on in the interface region of any given composite system under a given set of conditions. Wettability of the fiber by the matrix and the type of bonding between the two components constitute the primary considerations. Additionally, one should determine the characteristics of the interface and how they are affected by temperature, diffusion, residual stresses, and so on. We discuss below some of the interfacial characteristics and the associated problems in composites in a general way. The details regarding interfaces in polymer matrix, metal matrix, and ceramic matrix composites are given in specific chapters devoted to those composite types.

4.3 Interactions at the Interface

We mentioned above that interfaces are bidimensional regions. An initially planar interface, however, can become an interfacial zone having multiple interfaces resulting from the formation of different intermetallic compounds, interdiffusion, and so on. In such a case, besides the compositional parameter, we need other parameters to characterize the interfacial zone: for example, geometry and dimensions; microstructure and morphology; and mechanical, physical, chemical, and thermal characteristics of different phases present in the interfacial zone.

It commonly occurs that initially the components of a composite system are chosen on the basis of their mechanical and physical characteristics in isolation. It is important to remember, however, that when one puts together two components to make a composite, the composite will rarely be a system in thermodynamic equilibrium. More often then not, there will be present a driving force for some kind of interfacial reaction(s) between the two components leading to a state of thermodynamic equilibrium for the composite system. Of course, thermodynamic information such as phase diagrams can help predict the final equilibrium state of the composite. Data regarding reaction kinetics, for example, diffusivities of one constituent in another, can provide information about the rate at which the system would tend to attain the equilibrium state. In the absence of thermodynamic and kinetic data, experimental studies would have to be done to determine the compatibility of the components.

Quite frequently, the very process of fabrication of a composite can involve interfacial interactions that can cause changes in the constituent properties and/or interface structure. For example, if the fabrication process involves cooling from high temperatures to ambient temperature, then the difference in the expansion coefficients of the two components can give rise to thermal stresses of such a magnitude that the softer component (generally the matrix) will deform plastically. Chawla and Metzger [3] observed in a tungsten reinforced copper matrix (nonreacting components) that liquid copper infiltration of tungsten fibers at about 1100°C followed by cooling to room temperature resulted in a dislocation density in the copper matrix that was much higher in the vicinity of the interface than away from the interface. The high dislocation density in the matrix near the interface occurred because of plastic deformation of the matrix caused by high thermal stresses near the interface. Arsenault and coworkers [4, 5] found similar results in the system SiC whiskers in an aluminum matrix.

If powder metallurgy fabrication techniques are used, the nature of the powder surface will influence the interfacial interactions. For example, an oxide film on the powder surface affects its chemical nature. Topographic characteristics of the components can also affect the degree of atomic contact obtainable between the components. This can result in geometrical irregularities (e.g., asperities and voids) at the interface, which can be sources of stress concentrations.

4.4 Types of Bonding at the Interface

It is important to be able to control the degree of bonding between the matrix and the reinforcement. To do so, it is necessary to understand all the different possible bonding types, one or more of which may be acting at any given instant. We can conveniently classify the important types of interfacial bonding as follows:

1. Mechanical bonding
2. Chemical bonding
    (a) Dissolution and wettability bonding
    (b) Reaction bonding

Mechanical Bonding

Simple mechanical keying effects between two surfaces can lead to a considerable degree of bonding. Any contraction of the matrix onto a central fiber would result in a gripping of the latter by the former. There has been some work [6, 7] on metallic wires in metal matrices indicating that in the presence of internal compressive forces, a wetting or metallurgical bond is not quite necessary because the mechanical gripping of the fibers by the matrix is sufficient to cause an effective reinforcement as indicated by the occurrence of multiple necking in fibers. Hill et al. [8] have confirmed the mechanical bonding effects in tungsten wire/aluminum matrix composites. Chawla and Metzger [9] studied bonding between an aluminum substrate and anodized alumina (Al_2O_3) films and found that with a rough interface there occurred a more efficient load transfer from the aluminum matrix to the alumina.

Pure mechanical bonding alone will not be enough in most cases. However, mechanical bonding could add, in the presence of reaction bonding, to the overall bonding. Also, mechanical bonding is efficient in load transfer when the applied force is parallel to the interface.

Chemical Bonding

There are two types of chemical bonding:
1. *Dissolution and Wettability Bonding.* In this case, interaction between components occurs at an electronic scale. Since these interactions are of rather short range, it is important that the components come into an intimate contact on an atomic scale. This implies that surfaces should be appropriately treated to remove any impurities. Any contamination of fiber surfaces, or entrapped air or gas bubbles at the interface, will hinder the required intimate contact between the components.
2. *Reaction Bonding*. In this case, a transport of atoms occurs from one or both of the components to the reaction site, that is, the interface. This atomic transport is controlled by diffusional processes. Two polymer surfaces may form a bond owing to the diffusion of matrix molecules to the molecular network of the fiber, thus forming tangled molecular bonds at the interface. Coupling agents (silanes are the most common ones) are used for glass fibers in resin matrices. It is thought [10] that hydrogen bonds from between the silanol groups of glass fiber and partially

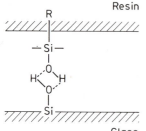

Resin

Glass

Fig. 4.3. Model of chemical bonding between glass fiber and resin through a silane coupling agent

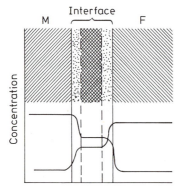

▤ Solid solutions
▨ Intermetallics
M   Metal
F   Fiber

Fig. 4.4. Interface zone in a metal matrix composite showing solid solution and intermetallic compound formation

hydrolyzed silanes as shown in Fig. 4.3. R is a resin-compatible group. A number of surface treatments (oxidative or nonoxidative) is available for carbon fibers to be used in polymeric materials. We describe these in Chap. 8. In metallic systems, one commonly finds solid solution and intermetallic compound formation. A schematic of the diffusion phenomenon between a fiber and a matrix resulting in solid solution as well as a layer of an intermetallic compound, M_xF_y, is shown in Fig. 4.4. The plateau region, which has a constant proportion of the two atomic species, is the region of intermetallic compound formation. The reaction products as well the reaction rates can vary depending on the matrix composition, reaction time, and temperature. Generally, one tries to fit such data to expressions of the form

$$x^2 = Dt \tag{4.4}$$

and

$$D = A \exp\left(-\frac{Q}{kT}\right) \tag{4.5}$$

where x is the interface reaction zone thickness (i.e., the diffusion distance), D is the diffusivity, t is the time, Q is the activation energy, k is the Boltzmann constant, T is the temperature in kelvin, and the preexponential constant A depends on the matrix alloy composition, fiber, and the atmosphere. Metcalfe [11] has described in detail the physical–chemical aspects of the interface in metallic composite systems, while Plueddemann [12] has described them in polymer composites.

4.5 Tests for Measuring Interfacial Strength

Numerous tests have been devised for characterizing the fiber/matrix interface strength. We now briefly describe some of these.

Single Fiber Tests

There exist many variants of this. Essentially, a portion of a single fiber is embedded in a matrix and the stress necessary to pull the fiber out is measured. A curved neck specimen configuration has been used to induce tensile debonding failure rather that a shear failure (13). A compressive axial load on such a curved neck specimen would result in transverse stresses, that is, perpendicular to the interface, owing to the different Poisson ratios of the matrix and the fiber.

Three-Point Bend Tests

Again there are many variants of this type of test. In one, with fibers running perpendicular to the specimen length, we measure the transverse strength of the composite. The following expression gives the stress σ in the outermost surface of the beam specimen:

$$\sigma = \frac{3}{2}\left(\frac{PS}{bh^2}\right)$$

where P is the applied load, b is the width, h is the height, and S is the span of the bend specimen. In another variant of this test, called the *interlaminar shear strength* (ILSS) *test*, the fibers run parallel to the length of the three-point specimen. The shear stress on the midplane is given by

$$\tau = \frac{3}{4}\left(\frac{P}{bh}\right)$$

where the symbols have the meaning given earlier. The maximum tensile stress, parallel to the fibers, occurs on the outermost surface and is given by

$$\sigma = \frac{3}{2}\left(\frac{PS}{bh^2}\right)$$

The ratio τ/σ ($= h/2S$) depends on the specimen geometry. A short-beam specimen is more likely to fail in shear. That is why the ILSS is also called the short-beam bend test. The ILSS test value is frequently used as a measure of the interfacial bond strength. It should be added that the experimental values generally depend on the fiber volume fraction in the composite; that is, they are not true indicators of the bond strength.

Indentation Method

An indentation test has been devised to measure the frictional strength of the interface in a ceramic matrix composite [14]. A standard microindentation hardness tester, for example, the Vickers pyramid, can be used. The indentor is loaded onto the fiber center in a section polished normal to the fiber axis. The applied load causes sliding along the fiber/matrix interface, which is manifested as a certain amount of depression of the fiber surface. Marshall [14] gives relationships for different geometries that allow an estimate of the frictional strength of the interface to be made. The test is especially suited for ceramic matrix composites.

References

1   J.L. Walter, H.E. Cline, and E. Koch, *Trans. AIME*, **245**, 2073 (1969).
2   R.E. Baier, E.G. Sharfin, and W.A. Zisman, *Science*, **162**, 1360 (1968).
3   K.K. Chawla and M. Metzger, *J. Mater. Sci.*, **7**, 34 (1972).
4   R.J. Arsenault and R.M. Fisher, *Scripta Met.*, **17**, 67 (1983).
5   M. Vogelsang, R.J. Arsenault, and R.M. Fisher, *Met. Trans. A*, **17**, 379 (1986).
6   R.M. Vennett, S.M. Wolf, and A.P. Levitt, *Met. Trans.*, **1**, 1569 (1970).
7   C. Schoene and E. Scala, *Met. Trans.*, **1**, 3466 (1970).
8   R.G. Hill, R.P. Nelson, and C.L. Hellerich, in *Proceedings of the 16th Refractory Working Group Meeting*, Seattle, WA, October 1969.
9   K.K. Chawla and M. Metzger, in *Advances in Research on Strength and Fracture of Materials*, Vol. 3, Pergamon Press, New York, 1978, p. 1039.
10   P. Ehrburger and J.B. Donnet, *Philos. Trans. R. Soc. London*, **A294**, 495 (1980).
11   A.G. Metcalfe, in *Interfaces in Metal Matrix Composites*, Academic Press, New York, 1974, p. 65.
12   E.P. Plueddemann, in *Interfaces in Polymer Matrix Composites*, Academic Press, New York, 1974, p. 174.
13   L.J. Broutman, in *Interfaces in Composites*, ASTM STP No. 452, American Society of Testing & Materials, Philadelphia, 1969.
14   D.B. Marshall, *J. Am. Ceram. Soc.*, **67**, C259 (1984).

Suggested Reading

R.E. Baier, E.G. Shafrin, and W.A. Zisman, Adhesion: Mechanisms that Assist or impede It, *Science*, **162**, 1360 (1968).
A.K. Dhingra and S.G. Fishman (eds.), *Interfaces in Metal Matrix Composites*, TMS-AIME, Warrendale, PA., 1986.
A.G. Metcalfe (ed.), *Interfaces in Metal Matrix Composites* (vol. 1 of the series Composite Materials), Academic Press, New York, 1974.
E.P. Plueddemann (ed.), *Interfaces in Polymer Matrix Composites* (vol. 6 of the series Composite Materials), Academic Press, New York, 1974.

Part II

5. Polymer Matrix Composites

Polymer matrix composites (PMCs) have established themselves as engineering structural materials, not just as laboratory curiosities or the cheap stuff for making chairs and tables. This came about not only because of the introduction of high-performance fibers such as carbon, boron, and Kevlar, but also because of some new and improved matrix materials (see Chap. 3). Nevertheless, glass fiber reinforced polymer composites represent the largest class among PMCs. We discuss the carbon fiber reinforced polymer composites separately, because of their great importance, in Chap. 8. In this chapter we discuss polymer composite systems containing glass, Kevlar, and boron.

5.1 Fabrication of PMCs

Various techniques are used for making glass fiber PMCs. Hand lay-up is the simplest of all. Glass fibers are laid onto a mold by hand and the resin is sprayed on or brushed on. Frequently, resin and fibers (chopped) are sprayed together onto the mold. In both these cases the deposited layers are densified with rollers. It is common practice to use polyesters or epoxy thermosetting resins. Accelerators and catalysts are frequently used. Curing may be done at room temperature or at higher temperatures.

Filament winding [1, 2] is another very versatile technique in which continuous tow or roving is passed through a resin impregnation bath and wound over a rotating or stationary mandrel. A roving consists of thousands of individual filaments. Figure 5.1 shows a schematic of this process. The winding of roving can be polar or helical. In the former the fiber tows do not cross over, while in the latter they do. The fibers are of course laid on the mandrel in a helical fashion in both polar and helical windings, the helix angle depending on the shape of the object to be made. Successive layers are laid on at a constant or varying angle until the desired thickness is attained. Curing of the thermosetting resin is done at an elevated temperature and the mandrel is removed. Very large cylindrical (e.g., pipes) and spherical (e.g., for chemical storage) vessels are built by filament winding. Glass, carbon, and aramid fibers are routinely used with epoxy, polyester, and vinyl ester resins for producing filament wound shapes.

Continuous, aligned fiber composites in the form of sections (e.g., I or T beams and hollow sections) can be made by pultrusion; see Fig. 5.2. Fibers (glass, Kevlar,

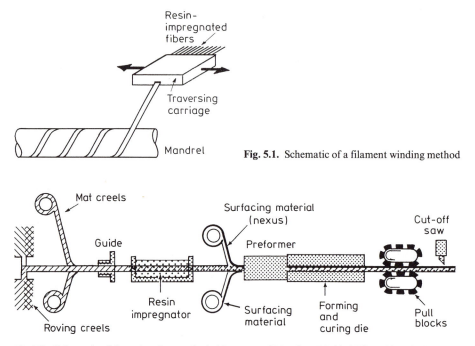

Fig. 5.1. Schematic of a filament winding method

Fig. 5.2. Schematic of the pultrusion method. (Courtesy of Morrison Molded Fiber Glass Co.)

or carbon), in desired orientations, are preimpregnated with thermosetting resin and pulled through a heated mold or die. Mold heating helps in curing the resin. Fiber placement, resin formulation and amount, catalyst level, die temperature, and pulling speed are critical parameters. Pulling speed may vary from a few centimeters to a few meters per minute. Meyer [3] provides the details regarding this process.

Bag molding processes [4] are used for making large parts. Vacuum or pressure bags containing fibers in predetermined orientations in a partially cured matrix (prepreg) are used in an autoclave for densification and curing of resin. Instead of prepregs, chopped fibers mixed with resin may also be used. An autoclave is a cylindrical oven capable of being pressurized. The bags are thin and flexible membranes made of rubber separate the fiber lay-ups from the compressing gases during curing of the resin. In bag molding, the bagged lay-ups are vented to pressures lower than those applied to the bag. Densification and curing are achieved by pressure differentials across the bag walls. In vacuum moldings, the bag contents are evacuated and atmospheric pressure consolidates the composite.

By far, the most common method of making high-performance PMCs involves the stacking of appropriately oriented prepregs (Fig. 5.3) and the subsequent cure of this composite laminate in an autoclave. By prepregs we mean thin sheets (~ 1 mm) of partially cured resin (sometimes called cured to stage B) containing unidirectionally aligned fibers. Generally, prepregs come with a backing paper that must be removed before laminating. More than one type of fiber may be used to produce so-called hybrid composites. A prepreg with fibers parallel to the long dimension is called a 0°lamina or ply. A prepreg that is cut with fibers perpendicular to the long

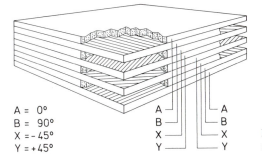

A = 0°
B = 90°
X = -45°
Y = +45°

Fig. **5.3.** Prepregs of different orientations stacked to form a laminate composite

dimension is designated as a 90°lamina, while a prepreg at an intermediate angle θ is designated as a θ ply. The exact orientation sequence is determined from elasticity theory (see Chap. 11) to give appropriate magnitudes and directions of stresses and to avoid unwanted twisting and/or torsion. This kind of laminate construction, mostly done by hand, is widely used in the aerospace industry. High fiber volume fractions can be obtained.

As pointed out in Sect. 3.1, thermoplastics soften on heating and therefore melt flow techniques of forming can be used. Such techniques include injection molding, extrusion, and thermoforming. Thermoforming involves the production of a sheet, which is heated and stamped, followed by vacuum or pressure forming. Generally, discontinuous fibrous (principally glass) reinforcement is used, which results in an increase of melt viscosity. Bader [5] has reviewed the topic of reinforced thermoplastics. Short fiber reinforced thermoplastic resin composites can also be produced by a method called *reinforced reaction injection molding* (RRIM) [6]. RRIM is actually an extension of the *reaction injection molding* (RIM) of polymers. In RIM, two liquid components are pumped at high speeds and pressures into a mixing head and then into a mold where the two components react to polymerize rapidly. An important example is a urethane RIM polymer. In RRIM, short fibers (or fillers) are added to one or both of the components. The equipment for RRIM must be able to handle rather abrasive slurries. The fiber lengths that can be handled are generally short owing to viscosity limitations. Because a certain minimum length of fiber called the critical length (see Chap. 10) is required for effective fiber reinforcement, more often than not RRIM additives are fillers rather than reinforcements. Most RIM and RRIM applications are in the automotive industry.

5.2 Structure and Properties of PMCs

Some microstructures of polymer matrices reinforced by continuous and discontinuous fibers are shown in Fig. 5.4. A transverse section of continuous glass fiber in an unsaturated polyester matrix is shown in Fig. 5.4a. The layer structure of an injection-molded composite consisting of short glass fibers in a semicrystalline polyethylene terephthalate (PET) is shown in Fig. 5.4b [7]. Essentially, it is a three-layer structure of different fiber orientations. More fibers are parallel to the mold fill

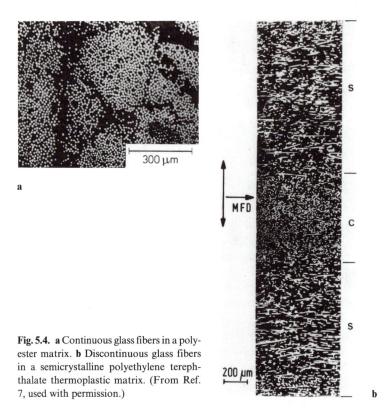

Fig. 5.4. a Continuous glass fibers in a polyester matrix. b Discontinuous glass fibers
in a semicrystalline polyethylene terephthalate thermoplastic matrix. (From Ref.
7, used with permission.)

direction (MFD) in the two surface layers, S, than in the transverse direction in the
central layer, C. Note the heterogeneity in the microstructure, which results in the
characteristically anisotropic behavior.

Laminates of polymer matrix composites made by the stacking of appropriately
oriented plies also result in composites with highly anisotropic characteristics. In
particular, the properties of continuous fiber reinforced polymers are quite a bit
higher in the longitudinal direction than in other directions. It turns out that
generally one finds the longitudinal properties of composites being quoted in the
literature for comparative purposes. The reader is warned that one must bear in
mind this discrepancy when comparing such data of highly anisotropic materials
with the data of isotropic materials such as common polycrystalline metals. Besides,
composites containing Kevlar fibers will not have as attractive properties in compression even in the longitudinal direction. A summary of some important characteristics of PMCs is presented in Table 5.1 [8].

The epoxy resins commonly employed as matrices in the composites meant for
use in the aerospace industry are fairly impervious to the range of fluids commonly
encountered, for example, jet fuel, hydraulic fluids, and lubricants [9]. There are,
however, two fundamental effects that must be taken into account when designing
components made of PMCs, namely, temperature and humidity. The combined
effect of these two, that is, hygrothermal effects, can result in a considerable
degradation in the mechanical characteristics of the PMCs. This is especially so in

Table 5.1. Representative properties of some PMCs[a]

Materials	Density (g cm⁻³)	Tensile modulus Longitudinal (GPa)	Transverse (GPa)	Shear modulus (GPa)	Tensile strength Longitudinal (MPa)	Transverse (MPa)	Compressive strength longitudinal (MPa)	Flexure modulus (GPa)	Flexure strength (MPa)	ILSS[b] (MPa)	Longitudinal coefficient of thermal expansion (10^{-6} K⁻¹)
Unidirectional E glass 60 v/o	2	40	10	4.5	780	28	480	35	840	40	4.5
Bidirectional E glass cloth 35 v/o	1.7	16.5	16.5	3	280	280	100	15	220	60	11
Chopped strand mat E glass 20 v/o	1.4	7	7	2.8	100	100	120	7	140	69	30
Boron 60 v/o	2.1	215	24.2	6.9	1400	63	1760	—	—	84	4.5
Kevlar 29 60 v/o	1.38	50	5	3	1350	—	238	51.7	535	44	—
Kevlar 49 60 v/o	1.38	76	5.6	2.8	1380	30	276	70	621	60	−2.3

[a] The values are only indicative and are based on epoxy matrix at room temperatures.
[b] ILSS, interlaminar shear strength.
Source: Adapted with permission from Ref. 8.

high-performance composites such as those used in the aerospace industry where dimensional tolerances are rather severe. Polymers absorb moisture by volumetric diffusion and in the absence of inorganic impurities, water merely causes a large amount of swelling and plasticizes the polymer; that is, the glass transition temperature of the polymer is. reduced [10]. The softening of epoxy matrix results in a reduction of the matrix support to the fibers. Such matrix support of fibers is more critical in compression loadings and to a slightly less extent in shear loadings [9]. If the polymer contains traces of salt, the polymer can act as a semipermeable membrane and the resulting osmotic pressure can cause cracking [10]. Thus, the effect may be reversible in unreinforced polymers free of inorganic impurities; but in fiber reinforced PMCs it can cause serious degradation of mechanical properties by interface delamination, microcracking, and so on [11]. In the case of glass fiber reinforced polymer matrix composites, it should be mentioned that bulk glass is known to be susceptible to stress corrosion in the presence of water or water vapor. The rate of environmental degradation generally depends on temperature, relative humidity, and diffusivity.

Degradation owing to ultraviolet radiation is another important environmental effect. Ultraviolet radiation breaks the covalent bonds in organic polymers. Sometimes a prolonged exposure of epoxy laminates to ultraviolet radiation results in a slight increase in strength, attributed to postcuring of the resin, followed by a gradual loss of strength as a result of laminate surface degradation [12].

Fracture in PMCs, as in other composites, is associated with the characteristics of the three entities: fibers, matrix, and interface. Specifically, fiber/matrix debonding, fiber pullout, fiber fracture, crazing and fracture of the matrix are the energy-absorbing phenomena that can contribute to the failure process of the composite. Of course, the debonding and pull out processes depend on the type of interface. At low temperatures, the fracture of a PMC involves a brittle failure of the polymeric matrix accompanied by pullout of the fibers transverse to the crack plane. Figure 5.5 shows this kind of fracture at $-80°C$ in the case of a short glass fiber/PET composite. Note the brittle fracture in the matrix. At room temperature, the same polymeric matrix (PET) deformed locally in a plastic manner, showing crazing [7]. Generally, stiffness and strength of a PMC increase with the amount of stiff and

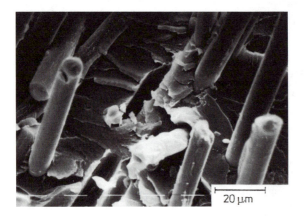

20 μm

Fig. 5.5. Brittle matrix fracture and fiber pullout in a short glass fiber reinforced polyethylene terephthalate (PET) composite fractured at $-80°C$. (From Ref. 7, used with permission.)

strong fibers introduced in a polymer matrix. The same cannot be said unequivocally for the fracture toughness. The toughness of the matrix and several microstructural factors related to the fibers and the fiber/matrix interface have a strong influence on the fracture toughness of the composite. Friedrich [7] has reviewed the fracture behavior of short fiber reinforced thermoplastic matrix composites. He describes the fracture toughness of the composite in an empirical manner by a relationship of the form

$$K_{cc} = M K_{cm}$$

where K_{cm} is the fracture toughness of the matrix and M is a microstructural efficiency factor. M can be larger than 1 and depends on fiber amount, fiber orientation, and the fiber orientation distribution over the fracture plane, as well as the deformation behavior of the matrix and the relative effectiveness of all the energy-absorbing mechanisms.

5.3 Interface in PMCs

Interfaces in glass fiber reinforced composites have been studied more extensively than in other systems. This is not surprising in view of the fact that glass fiber composites are the oldest composite structural materials. Glass fibers are given a "size" as soon as they come out of the bushings. The size is generally a water-based emulsion and its application allows one to handle glass fibers during composite manufacture without damaging their surface. Sometimes the size also contains adhesion promoters or coupling agents, while at other times, the size is removed and a coupling agent is applied to promote wetting and adhesion by the liquid resin. Commonly, the size contains polyvinyl acetate as a film former and either an organosilane or an organometallic complex as a coupling agent. Even though wetting of glass fibers by liquid resin is obtained through the coupling agent, one must realize that as the matrix cures it shrinks, and this can cause dewetting, cracking, and/or void formation at the interface. The differential shrinkage between fiber and matrix during fabrication can also cause residual stresses. This differential contraction can be advantageous if it causes the matrix to grip the fiber and leads to mechanical or frictional adhesion between the two.

There exists a good deal of controversy as to the exact functioning mechanisms of the coupling agents. Hull [13] gives a good description of the glass fiber/polyester resin interface. The glass surface has randomly distributed groups of various oxides; some of these, for example, SiO_2, Fe_2O_3, and Al_2O_3, absorb water as hydroxyl groups, —M—OH where M can be Si, Fe, and Al, and as molecular water attached to the (—M—OH) groups. If the oxides are of the hygroscopic variety, they become hydrated with water absorption. The important point is that glass absorbs water very readily to form a well-bonded surface layer, and avoidance of such water absorption is very difficult in practice. If the glass surface stays in contact with water for too long a time, hygroscopic elements can get dissolved and a porous surface consisting of nonhydrated oxides is left. In any event, strong bonding between fiber (oxide groups) and polymer molecules needs some kind of a chemical bond after

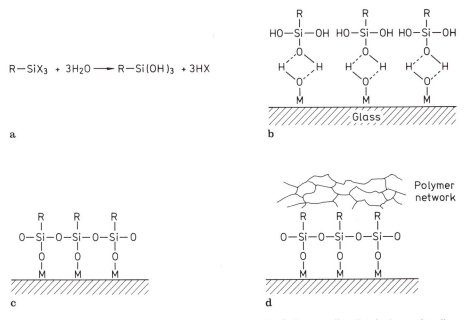

$$R{-}SiX_3 \;+\; 3H_2O \longrightarrow R{-}Si(OH)_3 \;+\; 3HX$$

a   b

c   d

Fig. 5.6. Function of a silane coupling agent: **a** hydrolysis of silane to silanol; **b** hydrogen bonding between hydroxyl groups of silanol and a glass surface; **c** polysiloxane bonded to a glass surface; and **d** resin-compatible functional groups R after reacting with polymer matrix. (From *An Introduction to Composite Materials*, by Derek Hull. Copyright © (1981), Cambridge University Press. Reprinted with permission.)

intimate molecular contact has been obtained by wetting. Examples of coupling agents commonly used are organometallic or organosilane complexes. It is thought by some researchers [14] that the coupling agents create a chemical bridge between the glass surface and the resin matrix. Other researchers [15] do not subscribe to this view.

The chemical bridge theory goes as follows. Silane molecules are multifunctional groups with a general chemical formula of $R{-}SiX_3$, where X stands for hydrolyzable groups bonded to Si. X can be, for example, an ethoxy group $-OC_2H_5$ and R is a resin-compatible group. They are hydrolyzed in aqueous size solutions to give trihydroxy silanols (Fig. 5.6a). These trihydroxy silanols get attached to hydroxyl groups at the glass surface by means of hydrogen bonding (Fig. 5.6b). During the drying of sized glass fibers, water is removed and a condensation reaction occurs between silanol and the glass surface and between adjacent silanol molecules on the glass surface, leading to a polysiloxane layer bonded to the glass surface (Fig. 5.6c). Now we can see that the silane coating is anchored at one end, through the R group, to the uncured epoxy or polyester matrix, and at the other end to the glass fiber through the hydrolyzed silanol groups. On curing, the functional groups R either react with the resin or join the resin molecular network (Fig. 5.6d).

Appealing though the above chemical bridge model of silane coupling is, there are certain shortcomings. The interface model shown in Fig. 5.6d will result in such a strong bond that it will fail at the very low strains encountered during the curing of

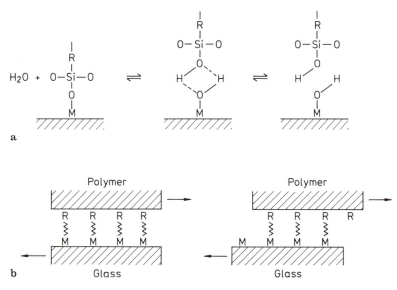

a

b

Fig. 5.7. a Plueddemann's reversible bond associated with hydrolysis. **b** Shear displacement at a glass/polymer interface without permanent bond rupture. (From *An Introduction to Composite Materials*, by Derek Hull. Copyright © (1981), Cambridge University Press. Reprinted with permission.)

the resin and resulting from differential thermal contraction. Also, under conditions of industrial application of silanes from aqueous solution, a covalent reaction to the glass fiber surface does not occur unless a primary or secondary amine is present [15, 16]. Plueddemann [17] has developed some ideas about the interaction of water and coupling agents at the interface in terms of a dynamic equilibrium. Plueddemann's model is shown in Fig. 5.7. In the presence of water at the interface (it can diffuse in from the resin), the covalent M—O bond hydrolyzes as shown in Fig. 5.7a. Now if there occurs a shear parallel to the interface, the polymer and glass fiber can glide past each other without a permanent bond rupture (Fig. 5.7b). Ishida and Koenig [18] used infrared spectroscopy to obtain experimental evidence for this reversible bond mechanism.

A Kevlar fiber surface bonds well with epoxy resins. For this reason, sometimes an epoxy-based "size" is applied to Kevlar fibers before incorporating them in other polymer matrices. A polyvinyl alcohol (PVA) size is sometimes given to protect a Kevlar fiber surface during handling. Wake [19] has proposed the reaction indicated in Fig. 5.8 between an isocyanate-linked polymer and Kevlar fiber. Epoxy-coated boron fibers and polyester-coated glass fibers are other examples of polymers used as coupling agents.

Interface modification is another important way of improving the properties of composite materials. Kardos and coworkers [15, 16] have modified the interface in PMCs by placing a ductile material or an intermediate modulus material at the interface. The objective in both cases is to reduce the stress concentration at the interface owing to a large modulus mismatch between fiber and matrix. The ductile interlayer provides a means of local deformation, thereby reducing interfacial stress concentrations.

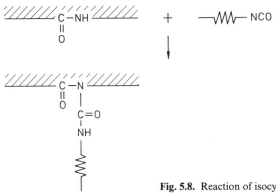

Fig. 5.8. Reaction of isocyanate-linked polymer with Kevlar fiber.

5.4 Applications

Glass fiber reinforced polymers are used in a wide variety of industries: from sporting goods to civil construction to aerospace. Tanks and vessels (pressure and nonpressure) in the chemical process industry, as well as process and effluent pipelines, are routinely made of glass fiber reinforced polyester resin. Figure 5.9 shows a wide variety of fiberglass/resin matrix structural shapes made by the pultrusion technique. S-2 glass fibers and Kevlar fibers are used in the storage bins and floorings of civilian aircraft. Other aircraft applications include doors, fairings, and radomes. Kevlar is also used in light load-bearing components in helicopters and small planes. In most applications involving glass fiber reinforced polymers, Kevlar fibers can be substituted for glass without much difficulty. Racing yachts and private boats are examples of Kevlar fiber making inroads into the glass fiber fields where performance is more important than cost. Drumsticks made with a pultruded core containing Kevlar fibers and a thermoplastic injection molded cover are shown in Fig. 5.10. These drumsticks last longer than the wooden ones, are lightweight, will not warp, and are more consistent than wooden sticks. Military

Fig. 5.9. A large variety of fiberglass/resin structural shapes made by pultrusion are available. (Courtesy of Morrison Molded Fiber Glass Co.)

Fig. 5.10. Drumsticks made with a pultruded Kevlar core and a thermoplastic injection molded cover. (Courtesy of Morrison Molded Fiber Glass Co.)

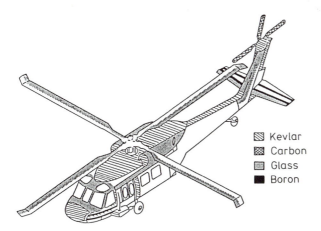

▨ Kevlar
▩ Carbon
▨ Glass
■ Boron

Fig. 5.11. Use of PMCs in a helicopter

applications vary from ordinary helmets to rocket engine cases. One has to guard against using Kevlar reinforcement in situations involving compressive, shear, or transverse tensile loading paths. in such situations, one should resort to hybrid composites (using more than one fiber species with Kevlar in the direction of longitudinal tension). In general, long-term and fatigue characteristics of Kevlar/epoxy composites are better than those of glass/epoxy composites. The fatigue resistance of Kevlar PMCs is inferior to that of carbon PMCs. Use of PMCs in military and commercial helicopters has been most impressive. Figure 5.11 shows the use of PMCs, including carbon fiber reinforced polymers, in the UH-60A Black Hawk military helicopter. Note the extensive use of Kevlar and glass fiber reinforced epoxy composites. The main driving force in such applications is of course that of weight reduction. An almost all composite airplane, shown in the frontispiece, a Beechcraft product, gives a pretty good idea of how much PMCs have become standard and routine engineering structural materials.

Boron/epoxy composites are used in sporting goods (e.g., golf clubs and tennis rackets), in the horizontal stabilizers and tail sections of military aircraft, in the foreflap of the Boeing 707, and so on. The high cost of boron fibers will keep them restricted to specialty items in the aerospace field.

An example of the use of newer thermoplastic polymeric matrix materials is shown in Fig. 5.12. Unidirectional glass fiber prepreg tape using Ryton polypheny-

Fig. 5.12. Unidirectional glass fiber prepreg tape in a Ryton polyphenylene sulfide resin matrix. The larger tape is about 10 cm wide. (Courtesy of Phillips Petroleum.)

lene sulfide resin as a matrix is available from Phillips Petroleum Co. Possible fields of application include automotive, chemical process, and aerospace industries.

References

1 A.M. Shibley, in *Handbook of Composite Materials*, Van Nostrand Reinhold, New York, 1982, p. 448.

2 Yu.M. Tarnopol'skii and A.I. Bail', in *Fabrication of Composites*, North-Holland, Amsterdam, 1983, p. 45.

3 R.W. Meyer, *Handbook of Pultrusion Technology*, Chapman & Hall, New York, 1985.

4 A. Slobodzinsky, in *Handbook of Composite Materials*, Van Nostrand Reinhold, New York, 1982, p. 368.

5 M.G. Bader, in *Fabrication of Composite Materials*, North-Holland, Amsterdam, 1983, p. 177.

6 R.J. Lockwood and L.M. Alberino, in *Advances in Urethane Science and Technology*, vol. 8, Technomic Press, Westport, CT, 1981.

7 K. Friedrich, *Composites Sci. Tech.*, **22**, 43 (1985).

8 N.L. Hancox, in *Fabrication of Composite Materials*, North-Holland, Amsterdam, 1983, p. 1.

9 B.W. Anderson, in *Advances in Fracture Research*, ICF6, New Delhi, Pergamon Press, Oxford, 1984, vol. 1, p. 607.

10 K.H.G. Ashbee, F.C. Frank, and R.C. Wyatt, *Proc. R. Soc.*, **A300**, 415 (1967).

11 G.S. Springer (ed.), *Environmental Effects on Composite Materials*, Technomic Press, Westport, CT, 1981.

12 H.W. Bergmann, in *Advances in Fracture Research*, ICF6, New Delhi, Pergamon Press, Oxford, 1984, vol. 1, p. 569.

13 D. Hull, *An Introduction to Composite Materials*, Cambridge University Press, Cambridge, U.K. 1981, p. 42.

14 C.E. Knox, in *Handbook of Composite Materials*, Van Nostrand Reinhold, New York, 1982, p. 136.

15 J.L. Kardos, in *Molecular Characterization of Composite Interfaces*, Plenum Press, New York, 1985, p. 1.

16 R.L. Kaas and J.L. Kardos, *Polym. Eng. Sci.*, **11**, 11 (1971).

17 E.P. Plueddemann, in *Interfaces in Polymer Matrix Composites*, Academic Press, New York, 1974, p. 174.

18 H. Ishida and J.L. Koenig, *J. Colloid Interface Sci.*, **64**, 555 (1978).

19 W.C. Wake, *J. Adhesion*, **3**, 315 (1972).

Suggested Reading

P. Ehrburger and J.B. Donnet, Interface in Composite Materials, *Philos. Trans. R. Soc. London*, **A294**, 495 (1980).

A. Kelly and S.T. Mileiko (eds.), *Fabrication of Composites*, North-Holland, Amsterdam, 1983.

G. Lubin (ed.), *Handbook of Composites*, Van Nostrand Reinhold, New York, 1982.

B.R. Noton (ed.), *Engineering Applications of Composites*, Academic Press, New York, 1974.

E.P. Plueddemann (ed.), *Interfaces in Polymer Matrix Composites*, Academic Press, New York, 1974.

M.M. Schwartz, *Composite Materials Handbook*, McGraw-Hill, New York, 1984.

6. Metal Matrix Composites

The boron fiber reinforced 6061 aluminum matrix composite system was developed in the 1960s. Unidirectionally solidified eutectics with an aligned two-phase microstructure were produced about the same time. Carbon fiber reinforced metallic composites were successfully made in the 1970s. With the availability of a wide variety of SiC and Al_2O_3 reinforcements, the research activity in the area of metal matrix composites increased tremendously the world over. Among the important MMC systems, we can include the following:

1. Boron/aluminum
2. Carbon/aluminum
3. Al_2O_3/Al and Al_2O_3/Mg
4. SiC/Al
5. Eutectic or in situ composites (really a subclass of MMCs)

These and other MMC systems afford us high specific strength and specific modulus plus a service temperature capability much higher than that of polymer matrix composites. Potentially, the excellent toughness and good environmental resistance of metallic matrices can result in quite superior MMC products. In this chapter we describe the fabrication, properties, interface characteristics, and applications of some important MMCs.

6.1 Fabrication of MMCs

New fabrication methods were developed in the 1970s and 1980s. Although MMCs can be finished by secondary processing methods, it appears that improved processing leading to a near net shape composite component in a cost effective manner will be the key to their commercial success. Cornie et al. [1] have reviewed the processing of metal and ceramic matrix composites.

A simple classification of metal matrix composite fabrication methods can be written as follows:

1. Solid state fabrication techniques
2. Liquid state fabrication techniques
3. In situ fabrication techniques

In 1 and 2, it may be necessary to coat the fiber or prepare fiber/matrix preforms for later consolidation and/or secondary fabrication. Thus, it would be in order to

250 μm

Fig. 6.1. Transverse section of SiC/Al wire preform

describe briefly some of the fiber surface preparation and preforming techniques before describing the composite fabrication processes.

Fiber surfaces may be coated with suitable materials to (a) improve wettability and adhesion and (b) prevent any adverse chemical interaction between the fiber and matrix at elevated temperatures. Boron fibers having a silicon carbide or boron carbide coating for use in an aluminum or titanium matrix are a good example. Boron fibers with a SiC coating (trade name Borsic) are produced by chemical vapor deposition (CVD) of SiC from a mixture of hydrogen and methyldichlorosilane to a thickness of about 2–3 μm. The Borsic fiber is especially suited for aluminum matrix composites involving high-temperature fabrication or applications at temperatures greater than 300°C. Boron carbide coated boron fibers (coating thickness about 5 μm) are more suited for titanium alloy matrices. Boron carbide coating is amorphous and is deposited by CVD [2].

In the case of carbon fibers for use in an aluminum matrix, titanium and boron are codeposited by CVD on carbon fibers. Deposition of Ti/B is obtained by reduction of $TiCl_4$ and BCl_3 by zinc vapors [3, 4]. The coating thickness formed at 650°C in 6 min is about 20 nm and is substantially TiB_2. Bundles of Ti/B coated wires, called precursor wires, are passed through a molten aluminum bath to make preforms for further consolidation into a C/Al composite component. The preforms can also be in the form of sheets. Similar wire or sheet preforms can be made for SiC/Al for posterior consolidation. Figure 6.1 shows a transverse section of such a SiC/Al wire preform.

6.1.1 Solid State Fabrication

Alternate layers of properly spaced boron fibers and aluminum foils are stacked to make the desired fiber volume fraction and fiber orientation. A resin-based fugitive binder is used to keep the boron fibers in place. A combination of heat and pressure, in vacuum, causes the matrix to flow around the fibers and make a bond with the next matrix layer, enclosing and gripping the fibers in between; see Fig. 6.2. B/Al composites are commercially consolidated in this fashion, under pressure and at moderately high temperatures. The selected temperature and the time of residence are not very high, so barrier coatings on the fibers are not necessary. Kreider and

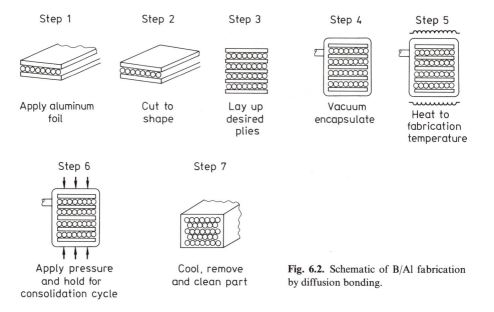

Step 1	Step 2	Step 3	Step 4	Step 5
Apply aluminum foil	Cut to shape	Lay up desired plies	Vacuum encapsulate	Heat to fabrication temperature

Step 6	Step 7	
Apply pressure and hold for consolidation cycle	Cool, remove and clean part	**Fig. 6.2.** Schematic of B/Al fabrication by diffusion bonding.

Prewo [5] give details regarding B/Al fabrication. Vacuum hot pressing is also used for B/Ti composites. Filaments are aligned and spaced over titanium sheets or foils by filament winding. Use is made of a fugitive binder to hold the fibers in place. Alternately stacked titanium sheets and boron fibers are sealed in stainless steel cans and subjected to hot pressing under vacuum in a manner similar to the one described above for B/Al [6].

Powder metallurgy techniques can profitably be used with discontinuous fibers or whiskers. Matrix metal powder and fibers are mixed and then pressed to consolidate. This may be followed by sintering to attain the theoretical density of the matrix.

Coextrusion or drawing can also be used to incorporate continuous ductile wires or filaments in a ductile metallic matrix. This method is employed in the commercial production of multifilamentary superconducting composites, which are discussed in Chap. 9.

Plasma spray, chemical or physical vapor deposition of matrix material onto properly laid-up fibers, followed by some kind of consolidation technique (e.g., vacuum hot pressing or hot isostatic pressing) are other solid state fabrication techniques for MMCs. Plasma spray is a very versatile technique of joining any two dissimilar materials. In particular, a low-pressure plasma deposition (LPPD) has been found to provide a very flexible method of producing composites with different volume fractions of reinforcement and matrix phase distribution, bonding, and so on [7]. Figure 6.3 shows a schematic of a LPPD process. A dc arc discharge through a gas mixture (Ar or N_2 plus 2–15% H_2 or He) is used to form a plasma (10,000–20,000 K) in the gun interior. Powders of the desired composition are injected into the plasma jet. The powder, entrained in the high-velocity plasma jet, is melted, transported, and impacted onto a substrate. The molten droplets solidify at 10^5–10^6

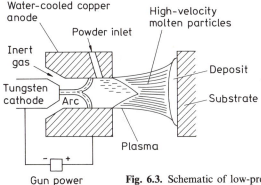

Water-cooled copper anode

Powder inlet

High-velocity molten particles

Inert gas

Deposit

Tungsten cathode

Arc

Substrate

Plasma

Gun power supply

Fig. 6.3. Schematic of low-pressure plasma deposition (LPPD) method. (From Ref. 7, used with permission.)

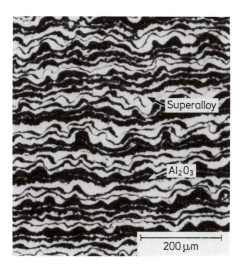

Superalloy

Al_2O_3

200 μm

Fig. 6.4. A ductile–brittle, poorly bonded continuous laminate composite consisting of 50% superalloy and 50% Al_2O_3. (From Ref. 7, used with permission.)

K s^{-1}. Discontinuous laminate composite structures may be obtained by spraying a powder blend of two or more different compositions. Continuous laminate structures can be obtained by using two plasma guns that traverse across a rotating cylindrical mandrel and deposit alternately the matrix and reinforcement materials. Ductile matrix–ductile reinforcement (e.g., superalloy–molybdenum) or ductile matrix–brittle reinforcement (e.g., superalloy–carbide) structures can be obtained. The degree of bonding can be controlled. Figure 6.4 shows a ductile–brittle, poorly bonded continuous laminate composite structure consisting of 50% superalloy–50% Al_2O_3. Good bonding is not obtained in this system mainly because of the low solubility of refractory oxides in the metal. Roll bonding [8] and explosive bonding [9] are also used to join two dissimilar metals.

6.1.2 Liquid State Fabrication

Basically, all methods in this category involve liquid metal (matrix) infiltration of fibers or fiber preforms. The infiltration may be carried out under atmospheric or inert gas pressure or under vacuum. In most of these techniques, the long continuous fibers must be properly aligned and distributed before infiltration by the matrix. Discontinuous fibers must be properly stirred and mixed with the molten metal. The liquid penetration around the fiber bundles can occur by capillary action, vacuum infiltration, or pressure infiltration. Clearly, the molten metal must wet the fiber surface. One specialized version of liquid state fabrication is called *squeeze casting* and is described below in detail.

Squeeze Casting

This technique involves the application of high pressures to the liquid metal during solidification. Donomoto et al. [10] and Toaz and Smalc [11] have given a description of the technique as used in diesel piston manufacture. Use of MMCs in diesel piston manufacture is illustrative of the advantage of MMCs. Squeeze casting is really an old process, also called *liquid metal forging* in earlier versions, and was developed to obtain pore-free, fine-grained aluminum alloy components having superior properties over those of conventional permanent mold casting. In particular, the process has been used in the case of aluminum alloys that are difficult to cast by conventional methods, for example, silicon-free alloys used in diesel engine pistons where high-temperature strength is required. Inserts of nickel containing cast iron, called Ni-resist, in the upper groove area of pistons have also been produced by the squeeze casting technique to provide wear resistance. Use of ceramic fiber reinforced metal matrix composites at locations of high wear and high thermal stress has resulted in a product much superior than the Ni-resist cast iron inserts.

The squeeze casting technique is shown in Fig. 6.5. A porous fiber preform (generally of discontinuous Saffil type Al_2O_3 fibers) is inserted into the die. Molten metal (aluminum) is poured into the preheated die located on the bed of a hydraulic press. The applied pressure (70–100 MPa) makes the molten aluminum penetrate the fiber preform and bond the fibers. Fiber volume fraction and orientation are important parameters in controlling the resistance against wear and seizure or

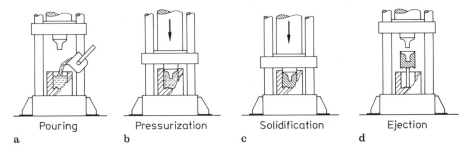

| Pouring | Pressurization | Solidification | Ejection |
| a | b | c | d |

Fig. 6.5a–d. Squeeze casting technique of composite fabrication

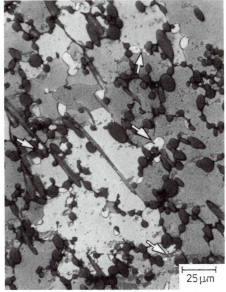

Fig. 6.6. a A cutout of a squeeze cast piston with MMC inserts indicated by dotted lines. (Courtesy of Toyota Motor Co.) **b** Microstructure of a squeeze cast Saffil (Al$_2$O$_3$) fiber reinforced aluminum alloy. Note the precipitates (indicated by arrows) nucleated preferentially at the fibers. (Courtesy of Warren Hunt, Alcoa Center.)

galling. Solidification of the matrix is completed under pressure, followed by ejection of the piston. It is worth pointing out that substitution of a Ni-Resist cast iron insert (heavy) with an alumina fiber reinforced aluminum matrix composite (light) can reduce the total piston weight by 5–10% [11]. In very-high-speed diesel engines, where inertial loads are critical, this is quite an advantage. Figure 6.6a shows a squeeze cast piston with a MMC insert while Fig. 6.6b shows the microstructure of a sequeeze cast Saffil (Al$_2$O$_3$)/Al composite. Note the preferential nucleation of the second-phase particles at the fiber/matrix interface. Squeeze cast MMCs are also being employed at the surface of a combustion bowl where thermal fatigue cracking is a problem.

6.1.3 In Situ Fabrication Techniques

Controlled unidirectional solidification of a eutectic alloy can result in a two-phase microstructure with one of the phases, present in lamellar or fiber form, distributed in the matrix. Figure 6.7 shows scanning electron micrographs of transverse sections of in situ composites obtained at different solidification rates [12]. The nickel alloy matrix has been etched away to reveal the TaC fibers. At low solidification rates, the TaC fibers are square in cross section, while at higher solidification rates, blades of TaC form. The number of fibers per square centimeter also increased with increasing solidification rate. Table 6.1 gives some important systems that have been investigated. A precast and homogenized rod of a eutectic composition is melted, in

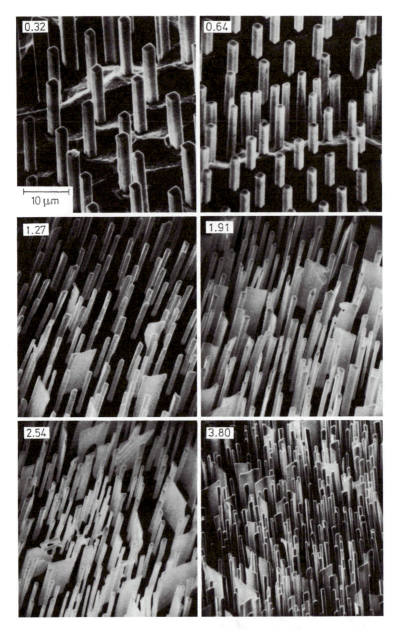

Fig. 6.7. Transverse sections of in situ composites obtained at different solidification rates. The nickel alloy matrix has been etched away to reveal the TaC fibers. (From Ref. 10, used with permission.)

Table 6.1. Some important in situ composite systems

System	Carbide (vol. %)	T_E[a] (°C)
Co-NbC	12	1365
Co-TiC	16	1360
Co-TaC	10	1402
Ni-HfC	15–28	1260
Ni-NbC	11	1330
Ni-TiC	7.5	1307

[a] T_E is the eutectic temperature.

vacuum or inert gas atmosphere. The rod is contained in a graphite crucible, which in turn is contained in a quartz tube. Heating is generally done by induction. The coil is moved up the quartz tube at a fixed rate. Thermal gradients can be increased by chilling the crucible just below the induction coil. Electron beam heating is also used, especially when reactive metals such as titanium are involved. From the mechanical properties point of view, a number of eutectic superalloy systems has been investigated in several research laboratories. The main advantages are improved rupture strength at high temperatures, low creep rates, and thermal stability of the microstructure. The reader is referred to the Suggested Reading list at the end of this chapter for further details; see especially the *Proceedings of the Conferences on In Situ Composites* (1973, 1976, 1979, 1982) and McLean (1983).

6.2 Interface in Metal Matrix Composites

In MMCs, as in other composites, we can have mechanical bonding as well as chemical bonding. We give a general account of the types of interfacial bonding in some important MMCs.

6.2.1 Mechanical Bonding

It was pointed out in Chap. 4 that a simple mechanical keying effect between two surfaces can lead to some bonding. Hill et al. [13] confirmed this experimentally for tungsten filaments in an aluminum matrix while Chawla and Metzger [14] observed mechanical gripping effects at Al_2O_3/Al interfaces. Hill et al. etched tungsten wires, original diameter 0.2 mm, to 0.165 mm along a portion of the wire and the rest was left at the original diameter. Then, about 12% of filaments were incorporated into an aluminum matrix by the liquid metal infiltration technique in vacuum. They evaluated three interface conditions by longitudinal tests of composites. The results are summarized in Table 6.2. In the case of smooth interface, there occurs a chemical reaction between the aluminum and the tungsten and the strength obtained is high, as expected. In the case of smooth interface with a graphite layer, the graphite barrier layer prevents the reaction from taking place, that is, no chemical bonding, and

Table 6.2. Effect of type of interface on strength

Interface condition	Percentage of theoretical strength
Smooth	95
Smooth, with a graphite barrier layer	35
Etched, with a graphite barrier layer	91

because the interface is smooth (no roughness), there is very little mechanical bonding either. The resultant strength is very low indeed. In the case of the etched wires with a graphite layer, there is no reaction bonding, but there is a mechanical keying effect because of the rough surface produced by etching. We note from Table 6.2 that mechanical bonding restores the strength to levels achieved in reaction bonding.

6.2.2 Chemical Bonding

Most metal matrix composite systems are nonequilibrium systems in the thermo-dynamic sense; that is, there exists a chemical potential gradient across the fiber/matrix interface. This means that given favorable kinetic conditions (which in practice means a high enough temperature), diffusion and/or chemical reactions will occur between the components. The interface layer(s) formed because of such a reaction will generally have characteristics different from those of either one of the components. At times, some controlled amount of reaction at the interface may even be desirable for obtaining strong bonding between the fiber and the matrix. Too thick an interaction zone, however, will adversely affect the composite properties.

Improvements in interfacial bonding in MMCs are frequently obtained by two methods: fiber surface treatment and matrix modification. We describe both of these techniques in some detail below.

Fiber Surface Treatment

Fiber surfaces may be coated with suitable materials to (a) improve wettability and adhesion and (b) prevent any adverse chemical interaction between the fiber and the matrix at elevated temperatures. Coatings of SiC and B_4C have been tried on boron fibers for use in metal matrices. Boron fibers with a SiC coating are more suited for an aluminum matrix while boron fibers with a B_4C coating are more appropriate for titanium matrices. Figure 6.8 plots the square of the reaction zone thickness against time for uncoated boron. B/SiC, and B/B_4C in a titanium matrix [2]. Note that all three curves follow the $x^2 = Dt$ type relationship (see Chap. 4), indicating that the reaction kinetics are diffusion controlled. Note also that for a given temperature, at least in the case of a titanium matrix, a B_4C coating is the best barrier.

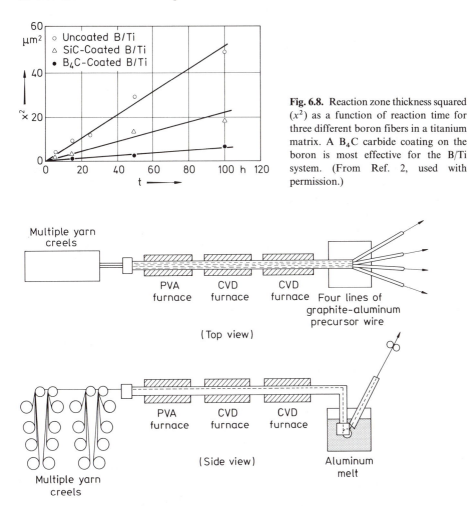

Fig. 6.8. Reaction zone thickness squared (x^2) as a function of reaction time for three different boron fibers in a titanium matrix. A B_4C carbide coating on the boron is most effective for the B/Ti system. (From Ref. 2, used with permission.)

Fig. 6.9. Schematic of the Ti/B codeposition process. (From Ref. 4, used with permission.)

In the case of carbon fibers in aluminum, poor wettability is a major problem. The wettability seems to improve somewhat with increasing temperature above 1000°C [15, 16], but it turns out that above 500°C a deleterious reaction occurs between the carbon fibers and the aluminum matrix, whereby Al_4C_3, a very brittle intermetallic compound, is known to form [17]. Fiber surface coatings have also been tried. Such coatings do allow wetting by low melting point metals but these coatings are also unstable in molten metals. Fiber degradation leading to reduced composite strength generally results. In the case of carbon fiber/aluminum composites, codeposition of titanium and boron onto carbon fibers before incorporating them into an aluminum matrix came to be established as a commercial method for carbon fiber surface treatment. Figure 6.9 shows a schematic of this process [4]. Multiple yarn creels result in increased capacity over single yarn creels. The polyvinyl alcohol (PVA) size on carbon fibers (meant for polymeric matrices; see Chap. 9) is removed in a furnace, designated as a PVA furnace in Fig. 6.9. The first CVD furnace in Fig. 6.9 is a precoat furnace for cleaning and activating the carbon yarn surface. The Ti/B coating is deposited in the second CVD furnace, followed by drawing through a molten aluminum bath.

Codeposition of titanium and boron on carbon fibers for use in magnesium matrix has also been tried. There is, however, a problem associated with this. The titanium-boron coating oxidizes very rapidly when exposed to air, and molten magnesium does not wet the oxidized coating. Thus, the coated carbon fibers must not be exposed to air which makes the processing somewhat difficult. This has led to work on a different type of coating, an air-stable coating of silicon dioxide which is readily wet by magnesium [18]. This coating is deposited on the fibers by passing them through an organometallic precursor solution, followed by hydrolysis or pyrolysis of the organometallic compound to form the silicon dioxide on the fiber surfaces. The presence of magnesium, silicon, and oxygen was confirmed by means of scanning auger and secondary ion mass spectroscopy. This mixed-oxide phase gives a graded interface with good bonding between the matrix and the fibers.

Matrix Modification

Modifying the matrix alloy composition to obtain adequate wetting and bonding between fiber and matrix is another route available in some systems. Remember that a decrease in fiber surface energy will promote wetting. Thus, the idea is to alter the matrix composition in such a way that dopants would react with the fiber in a controlled manner to give a thin fiber surface layer that will be wetted by the liquid matrix alloy. Wetting is rather poor between different forms of alumina and molten aluminum, with contact angles ranging from 180° (no wetting) at the melting point to greater than 60° at 1800 K. In particular, α-Al_2O_3 (FP)-based systems do involve the formation of spinels at the fiber/matrix interface which enhance the chemical bond at the interface [19]. Figure 6.10 shows a controlled amount of interfacial reaction at the Al_2O_3/Mg interface. Melt infiltrations of δ-alumina (Saffil) involving high pressure and rapid infiltration resulted in a minimal interfacial reaction, perhaps extending only to an atomic monolayer level [20]. However, a very good and successful example of the matrix modification approach to obtain enhanced wettability and bonding is that of alloying the aluminum matrix with lithium to promote wetting of the polycrystalline Al_2O_3 (FP) fibers [21]. Thus, aluminum with 2–3% lithium would wet the Al_2O_3 fibers in vacuum, thereby permitting the

1 μm

Fig. 6.10. A controlled amount of interfacial reaction at the Al_2O_3 fiber/Mg alloy matrix interface. M, F, and RZ denote matrix, fiber, and reaction zone. Darkfield (DF) transmission electron micrograph.

technique of metal infiltration of a fiber bundle to be used to prepare composites. Lithium concentration, process time, and temperature conditions are the important parameters controlling the interfacial reactions to obtain adequate bonding without fiber degradation or poor infiltration due to lack of wetting. It was observed [21] that lithium concentrations in excess of 3.5–4 wt.% significantly reduced the composite tensile strength. X-ray diffraction studies of Al_2O_3 fibers extracted from the matrix showed the presence of $LiAlO_2$ on the fiber surface [21]. The following chemical reaction has been proposed for the metallurgical bonding in FP fiber/Al-Li matrix [22]:

$$6\,Li + Al_2O_3 \rightarrow 3\,Li_2O + 2\,Al$$

$$Li_2O + Al_2O_3 \rightarrow 2\,LiAlO_2$$

Electron diffraction analysis showed the presence of the $Li_2O \cdot 5Al_2O_3$ spinel at the fiber surface [23]. In yet another study on this FP/Al-Li system [24], involving the effect of isothermal exposure for periods up to 100 h, two compounds, namely, α-$LiAlO_2$ and $LiAl_5O_8$, were identified in the interfacial reaction layer by electron diffraction.

The matrix modification approach has been tried for C/Al alloy matrix composites. Pepper et al. [25] obtained good compatibility between carbon fibers coated with Al-Si alloys when these fiber preforms were hot pressed. Kimura et al. [26] studied compatibility between carbon fibers and binary aluminum alloys containing indium, lead, and thallium and found considerable enhancement in the wetting of the carbon fibers.

Warren and Andersson [27] have reviewed the thermodynamics of chemical equilibria between SiC and some common metals. They divide these systems into *reactive* and *stable* types. In reactive systems, SiC reacts with the metal to form silicides and/or carbides and carbon. No two phase field exists in the ternary phase diagram showing SiC and the metal in equilibrium. For example, with nickel and titanium, the following reactions are possible:

$$SiC + Ni \rightarrow Ni_xSi_y + C$$

$$SiC + Ti \rightarrow Ti_xSi_y + TiC \quad \text{(low SiC fractions)}$$

Such reactions are thermodynamically possible between SiC and nickel or titanium, but, in practice, reaction kinetics determine the usefulness of the composite. In stable systems, SiC and a metallic matrix alloy can coexist thermodynamically, that is, a two phase field exists. Examples are SiC fibers in alloys of aluminum, gold, silver, copper, magnesium, lead, tin, and zinc. This does not imply that SiC will not be attacked. In fact, a fraction of SiC fibers in contact with molten aluminum can dissolve and react to give Al_4C_3:

$$SiC + Al \rightarrow SiC + (Si) + Al_4C_3$$

This reaction can happen because the section SiC-Al lies in a three phase field. Such reactions can be avoided by prior alloying of aluminum with silicon. Kohara [28] showed this beneficial effect of matrix alloying with the SiC/Al system. The addition of silicon to aluminum puts the overall composite composition in the two-phase Al-SiC field.

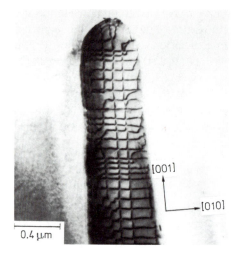

Fig. 6.11. A network of dislocations between the NiAl matrix and a chromium rod in a unidirectionally solidified NiAl-Cr eutectic. (Reprinted with permission from [Acta Met., 19, H.E. Cline, J.L. Walter, E.F. Koch, and L.M. Osika, 405], Copyright © [1971], Pergamon Press Ltd.)

6.2.3 Interfaces in In Situ Composites

Generally, the fiber/matrix interfaces in in situ composites have low energies. The spacing λ between rods or lamellae and the solidification rate R are related:

$$\lambda^2 R = \text{const.}$$

The composite strength depends on the interlamellar spacing λ according to a Hall–Petch equation (see Chap. 3):

$$\sigma = \sigma_0 + k\lambda^{-1/2}$$

where σ_0 is an intrinsic strength and k is a material constant.

X-ray and electron diffraction studies show preferred crystallographic growth directions, preferred orientation relationships between the phases, and low-index habit planes.

Boundaries between in situ components are usually semicoherent and the lattice mismatch across the interfaces can be accommodated by interface dislocations. Figure 6.11 shows a network of dislocations between the NiAl matrix and a chromium rod in a unidirectionally solidifed NiAl-Cr eutectic [29].

6.3 In Situ MMCs

Aligned eutectic alloys are potential candidates for jet engine blade applications. The NiTaC-based systems, having TaC as the reinforcing phase, have good mechanical properties and environmental resistance. Some of the important (from a mechanical property viewpoint) eutectic composite systems are given in Table 6.1. There are other systems, however, that are important from an electronic property viewpoint. van Suchtelen [30] has classified eutectic or in situ composites into two broad categories from an electronic property viewpoint:

1. *Combination Type Properties*. This can be further subdivided into (a) sum type and (b) product type. In sum type, properties of the constituent phases contribute proportionately to their amount. Examples are heat conduction, density, and elastic modulus. In the product type, the physical output of one phase serves as input for the other phase: for example, conversion of a magnetic signal into an electrical signal in a eutectic composite with one phase magnetostrictive and the other piezoelectric.

2. *Morphology-dependent Properties*. In this case the properties depend on the periodicity and anisotropy of the microstructure, the shape and size of the phases, and the amount of interface area between the phases. A good example is InSb-NiSb which is quasibinary system with a eutectic at 1.8% NiSb. Unidirectional solidification of a eutectic melt at a growth rate of 2 cm h^{-1} results in an aligned composite consisting of an InSb semiconducting matrix containing long hexagonal fibers of the NiSb phase. The magnetoresistance of the InSb-NiSb composite becomes extremely large if the directions of the metallic fibers, the electric current, and the magnetic field are mutually perpendicular. This characteristic has been exploited in making contactless control devices; a practical example is described in Sect. 6.6.

6.4 Discontinuous Reinforcement of MMCs

Among the MMCs having reinforcement in a discontinuous form, silicon carbide whiskers (SiC_w) and particles (SiC_p) in an aluminum matrix have attracted a good deal of attention. Aluminum matrix composites are of interest because they can be subjected to conventional metal working processes (e.g., rolling, forging, extrusion, machining, and swaging). Figure 6.12a shows a perspective montage (SEM) of a rolled plate of SiC (20 v/o)/2124 Al. The rolling direction is perpendicular to the original extrusion direction. The very small particles in Fig. 6.12a are the precipitates in the matrix after aging, while the large irregular particles are probably (Fe, Mn) Al_6. The distribution of SiC whiskers in the aluminum matrix as seen in a transmission electron microscope is shown in Fig. 6.12b, while a close up of the whisker/matrix interface region is shown in Fig. 6.12c. Note the waviness of the interface. Fu et al. [31] examined the interface chemistry and crystallography of C/Al and SiC_w/Al MMCs in the TEM. They observed that an oxide is present at some of the SiC/Al interfaces. Figures 6.13 a and b show the TEM bright and dark field micrographs of the interface region of 20 v/o SiC_w/Al. The crystalline γ-Al_2O_3 phase is about 30 nm. It was not uniformly present, however, at every interface. In C/Al composites, both fine-grained γ-Al_2O_3 and coarse-grained Al_4C_3 were found at the interfaces. On heat treating, some of the Al_4C_3 grew into and along the porous sites at the carbon fiber surface.

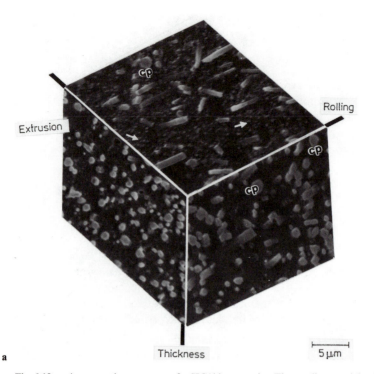

a

Fig. 6.12. a A perspective montage of a SIC/Al composite. The smallest particles (indicated by arrows on the top face) are precipitates from the matrix as a result of aging. Silicon carbide whiskers are the intermediate particles, both long and narrow and equiaxial ones having a diameter similar to the width of the long, narrow ones. The large, irregular particles are $(Fe, Mn)Al_6$ type constituent particles (some of these are marked cp). The liquid phase hot pressed billet was first extruded in the *extrusion* direction, forged along the *thickness* direction, and finally rolled along the *rolling* direction. (Courtesy of D.R. Williams and M.E. Fine.) **b** Distribution of SiC_w in an aluminum matrix (TEM). **c** A higher magnification of the whisker/matrix interface (TEM). (Courtesy of J.G. Greggi and P.K. Liaw.)

6.5 Properties

From a mechanical property point of view, DiCarlo [32] has summarized the requirements for strong and tough metal matrix composites. Generally, MMCs consist of a ductile matrix containing strong, brittle fibers. For maximum strength in a given direction, therefore, we need continuous fibers (i.e., high aspect ratio) strongly bonded to the matrix for efficient load transfer (more about this in Chap. 10). Fiber–matrix reactions at the interface during fabrication and/or in service and thermal stresses due to expansion mismatch between fiber and matrix are the possible sources of difficulties. For a tough MMC, we need (a) fibers with a high in situ strength and a low density of critical flaws and (b) maximum interfiber spacing for a given fiber volume fraction. Requirement (b) exploits the crack blunting characteristics of tough metals; this can be achieved for a given fiber volume fraction by using fibers with as large a diameter as possible. This is the driving force behind efforts to produce the large-diameter boron fibers mentioned in Sect. 2.3.

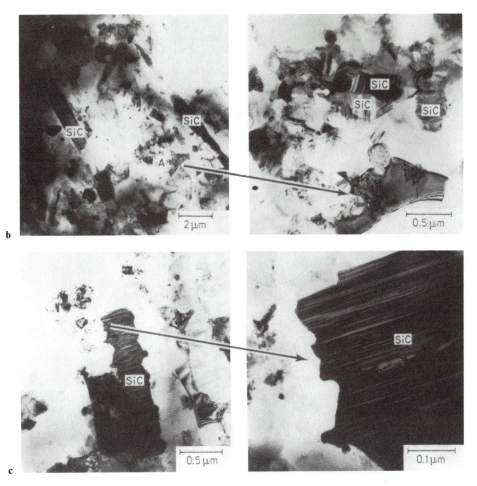

Fig. 6.12 b,c

As expected of fiber reinforced composites, the mechanical properties of MMCs are highly anisotropic. Figure 6.14 shows the stress–strain curves of B(10% V_f)/Mg for four different fiber orientations with respect to the loading directions [33]. Note that a very small amount of fibers at different orientations can make a tremendous difference in the stress–strain behavior. Keeping this property spread with orientation in mind, it is instructive to get some idea of the properties of some representative MMCs. Table 6.3 does this.

Generally, the strength properties vary linearly with fiber volume fraction. As examples, Figs. 6.15 a and b show the strength and modulus of Al_2O_3 (FP)/Al-Li composites as a function of fiber volume fraction [21].

Thermal stresses due to the differential in the expansion coefficients of fiber and matrix can attain unacceptable levels in MMCs. Among the various metal matrices, the thermal expansion coefficient of boron is close to that of titanium. Thus, the problem of thermal stresses will be minimized in this case. Chawla and Metzger [34] and Chawla [35–38] showed the importance of thermal stresses on microstructure,

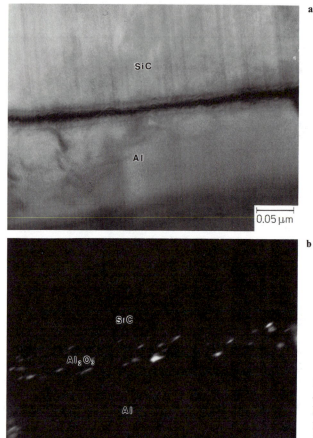

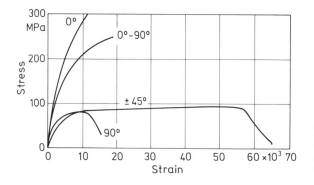

Fig. 6.13. Interface in SiC$_w$/Al composite: **a** bright field TEM, **b** dark field TEM showing the presence of Al$_2$O$_3$ at the interface. (From Ref. 31. Copyright ASTM. Reprinted with permission.)

Fig. 6.14. Stress–strain curves for B(10% V_f)/Mg for different fiber orientations. (From Ref. 33, used with permission.)

Table 6.3. Representative properties of some MMCs

Composite and direction	Fiber volume fraction (%)	Density (g cm^{-3})	σ_{max} (MPa)	E (GPa)
B/Al				
0°	50	2.65	1500	210
90°	50	2.65	140	150
SiC/Al				
0°	50	2.84	250	310
90°	50	2.84	105	—
SiC/Ti-6 Al-4V				
0°	35	3.86	1750	300
90°	35	3.86	410	—
Al$_2$O$_3$ (FP)/Al-Li				
0°	60	3.45	690	262
90°	60	3.45	172–207	152
C/Mg alloy (Thornel 50)	38	1.8	510	—
C/Al	30	2.45	690	160

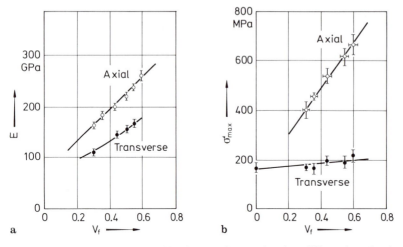

Fig. 6.15. Properties of Al$_2$O$_3$ (FP)/Al-Li composites as a function of fiber volume fraction (V_f): **a** axial and transverse Young's modulus versus fiber volume fraction, **b** axial and transverse ultimate tensile strength versus fiber volume fraction. (From Ref. 21, used with permission.)

strength, cracking, and general stress–strain behavior of tungsten/single-crystal copper matrix, tungsten/polycrystalline copper matrix, and B/Al and B/Mg composites. Thermal stresses in these systems, for a small change in temperature, can become large enough to deform the matrix plastically. Figure 6.10 shows the formation of twins in the matrix in the as-cast situation due to the thermal stresses generated on cooling. Similar results have been observed in SiC/Al [39, 40]. Thermal

cycling damage in metal matrix composites is somewhat attenuated if the maximum cycling temperature is above the recrystallization temperature of the metallic matrix [41, 42]. In such cases the work hardening in the matrix of the composite, resulting from plastic deformation caused by thermal stresses, is accompanied by structural recovery phenomena. Nieh and Karlak [43] showed that the presence of B_4C particles in 6061 Al matrix accelerated the aging response of the matrix. This was attributed to the high dislocation density, generated by the thermal mismatch between the reinforcement and the matrix, and the presence of a highly diffusive interface. Both factors led to an increase in the effective diffusivity of the precipitate forming alloy additions, which in turn led to an accelerated aging response of the matrix. Pedersen [44] has discussed the relationships between residual stresses in MMCs and the observed constraint effects. What is important to realize is that the matrix in fiber composites is not merely a kind of glue or cement to hold the fibers together [45]. The characteristics of the matrix, as modified by the introduction of fibers, must be evaluated and exploited to obtain an optimum set of properties of the composite. This realization is evidenced in the MMC field by the work cited above. In the field of PMCs, as noted in Chap. 5, efforts to modify the matrix characteristics are gaining importance. In the MMC area, some interesting work done at MIT has focused on matrix microstructural control in cast MMCs [46]. Not unexpectedly, it was found that the nature of the fibers, fiber diameter and distribution, as well as conventional solidification parameters, influenced the final matrix microstructure. Porosity is one of the major defects in cast MMCs owing to the shrinkage of the metallic matrix during solidification. At high fiber volume fractions, the flow of interdendritic liquid becomes difficult and large-scale movement of semisolid metal may not be possible. More importantly, the microstructure of the metallic matrix in a fiber composite can differ significantly from that of the unreinforced metal. Mortensen et al. [46] have shown that the presence of fibers influences the solidification of the matrix alloy. Figure 6.16a shows a cross section of an SCS-2 silicon carbide fiber/Al–4.5% Cu matrix. Note the normal dendritic cast structure in the unreinforced region, whereas in the reinforced region the dendritic morphology is controlled by the fiber distribution. Figure 6.16b shows the same system at

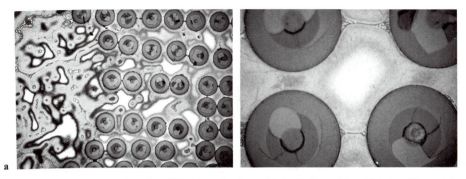

a   b

Fig. 6.16. Transverse section of an SCS-2 SiC fiber in an Al–4.5% Cu matrix. **a** Note the difference in the dendritic structure of the unreinforced and the fiber-rich regions of the matrix. **b** The second phase appears preferentially at the fiber/matrix interface or in the narrow interfiber region. The fiber diameter is 142 μm in both **a** and **b**. (Reprinted with permission from JOURNAL OF METALS, Vol. 37, No. 6, pgs 45, 47, and 48, a publication of The Metallurgical Society, Warrendale, Pennsylvania.)

a higher magnification. The second phase (theta) appears preferentially at the fiber/matrix interface or in the narrow interfiber spaces. This phenomenon can also be seen in the micrograph shown in Fig. 6.6b. Thus, it is possible to control the location of the second phase in the matrix, the amount of microsegregation, and the grain size of the matrix, as well as characteristics of the fiber/metal interface. Kohyama et al. [47] observed that shrinkage cavities present in the matrix in SiC/Al composites were the predominant crack initiation sites. They also observed a wavy interface structure between SiC and the aluminum matrix, indicating some mechanical bonding in addition to any chemical and physical bonding. In the system SiC/Mg (AZ19C), they observed a magnesium-rich interfacial layer that acted as the fracture-initiating site.

Superior high-temperature properties of MMCs have been demonstrated in a number of systems. For example, silicon carbide whiskers (SiC_w) improve the high-temperature properties of aluminum considerably. Figure 6.17 compares the elastic modulus, yield stress, and ultimate tensile strength of $SiC_w(21\%\ V_f)/2024$ Al composites to that of an unreinforced aluminum alloy [48]. Similar results have been obtained with continuous boron fibers coated with SiC (BORSIC) in an aluminum matrix [5]. The difference in favor of SiC is that one does not need any barrier coating at least up to the medium temperature range shown in Fig. 6.18. Note the

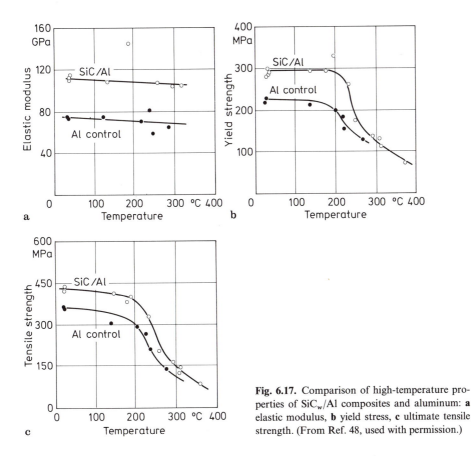

Fig. 6.17. Comparison of high-temperature properties of SiC_w/Al composites and aluminum: **a** elastic modulus, **b** yield stress, **c** ultimate tensile strength. (From Ref. 48, used with permission.)

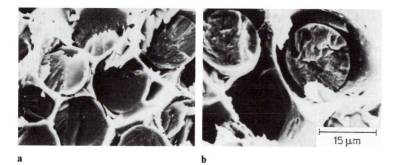

a   b

Fig. 6.18. Tensile fracture in SiC (Nicalon)/aluminum: **a** at room temperature showing a planar fracture, **b** at 500°C showing fiber/matrix separation and fiber pullout leaving a hole. (Courtesy of K. Okamura.)

more or less planar fracture with no fiber pullout at room temperature (Fig. 6.18a). At 500°C, there occurs a loss of adhesion between the fiber and the matrix with the accompanying fiber/matrix separation and fiber pullout. The fiber pullout has left a hole in the center; see Fig. 6.18b.

High-modulus carbon fiber/aluminum composites combine a very high stiffness with a very low thermal expansion due mainly to the almost zero longitudinal expansion coefficient of carbon fibers. Carbon/aluminum composites, however, are susceptible to galvanic corrosion between carbon and aluminum [49]. Carbon is cathodic in nature while aluminum is anodic. Thus, galvanic corrosion can be a serious problem in joining aluminum to a carbon fiber composite. A common solution is to have an insulating layer of glass between the aluminum and the carbon composite. Any welding or joining involving localized heating could also be a potential source of problems because aluminum carbide may form as a result of overheating, which will be detrimental to the mechanical and corrosion properties [49].

The creep behavior of MMCs has been investigated by a number of workers [50–52]. Kelly and Street [50] showed that the fiber aspect ratio and the matrix stress exponent in the power law creep relationship are important parameters. Lilholt [51] has expressed power law creep in terms of an effective rather than an applied stress. McLean [52] has discussed the modeling of creep behavior in MMCs. In MMCs where the differences in the melting points of the matrix and the reinforcing phase are small, for example, in a nickel-based superalloy matrix reinforced with refractory wires or in an in situ composite, both fiber and matrix creep at high temperatures. This is because the operating homologous temperature for the matrix and fiber is about the same. On the other hand, in MMCs consisting of ceramic fibers in a metal, for example, SiC/Al or Al_2O_3/Al, the operating homologous temperatures for fiber and matrix may be 0.1–0.3 and 0.4–0.7, respectively. McLean pointed out that in such composites the matrix is subject to creep rates orders of magnitude greater than those of the fibers, leading to a situation wherein we are likely to have elastically deforming fibers in a creeping matrix. Thus, it is unlikely that we will observe steady-state creep rates in such composites. Instead, the creep rates will progressively fall and asymptotically approach zero as the strain approaches an equilibrium value. In discontinuous fiber MMCs, steady-state creep is achieved because of matrix flow around the fiber ends. Off-axis loading can also result in steady-state creep because of nonaxial stress components. Thus, fiber misalignment is likely to degrade the composite creep performance. Indirect matrix strengthening can result because of the introduction of thresholds for matrix deformation or the suppression of glide creep in the matrix and its substitution by creep resulting from dislocation climb.

Toughness and fatigue are the other important properties where MMCs can show important benefits. Relatively less attention has been paid to the fatigue behavior of MMCs. The reason for this may be that fatigue has not been regarded as important a problem as, say, high-temperature strength by the users of composites. This may be understandable as far as space applications are concerned but for large-volume applications involving the use of MMCs in areas such as the automotive industry, the fatigue behavior of these materials must be investigated. Williams and Fine [53] investigated the fatigue behavior of SiC whisker reinforced 2124 Al alloy composites. They found that unbonded SiC_p and non-SiC intermetallics were the fatigue crack-initiation sites. The unbonded SiC particles occur when clusters of SiC are present. Thus, not unexpectedly, reducing the clustering of SiC and the number and size of the intermetallics resulted in increased fatigue life. Cyclic loading results in a uniform distribution of dislocations in the metal matrix [53, 54]. Figure 6.19 shows the dislocation distribution in an aluminum matrix before and after fatigue testing of a SiC_w/Al composite [53]. Note the uniform dislocation distribution after fatigue testing. Rosenkranz et al. [55, 56] have studied the low-cycle and high-cycle fatigue behavior of steel fiber reinforced silver matrix composites. They examined the influence of the fiber and matrix strength levels, the interfacial bond strength,

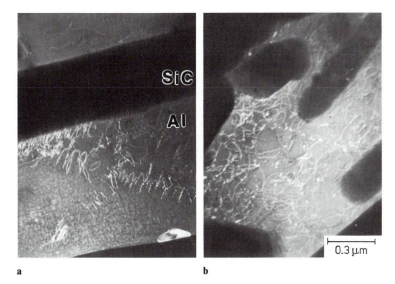

a   b

Fig. 6.19. Dislocation distribution in the aluminum matrix of a SiC_w/Al composite: **a** inhomogeneous dislocation distribution before testing, **b** uniform dislocation distribution after fatigue testing. (From Ref. 53, used with permission.)

and the mean fiber diameter. While the variations in the matrix and fiber strengths and the interfacial bond strength had significant influence on the fatigue damage and failure, the change of fiber diameter in the range of 35–200 µm had little effect. Logsdon and Liaw (57) compared the mechanical behavior of SiC$_w$ (20% V_f)/2124 Al(T6), SiC$_p$(25% V_f)/6061 Al(F and T6) and SiC$_w$(25% V_f)/6061 Al(T6) with that of the corresponding wrought aluminum alloys. The composites showed increased yield and ultimate strength but substantially lower ductility and fracture toughness, lower crack propagation resistance, and about the same values of threshold stress intensity range [57]. SiC/Al composites produced by standard metallurgical processes have generally superior mechanical properties than does the matrix alloy. The only drawback has been rather low tensile ductility and fracture toughness values of the composite vis-à-vis the matrix alloy [57, 58]. The room temperature fracture toughness values, K_{IC} varied between 7 and 14 MPa m$^{1/2}$ for SiC$_w$/Al, while those for SiC$_p$/Al varied between 12 and 20 MPa m$^{1/2}$. Crowe et al. [58] found the fracture toughness of SiC$_w$/Al composites to be dependent on the projected area of SiC in the fracture plane, leading to an orientation dependence of K_{IC}. Among the explanations for low toughness values of the composites are the following variables: the type of intermetallic particles, inhomogeneous internal stress, and silicon carbide distributions. There are indications that some success has been obtained in improving the fracture toughness of SiC$_w$/Al composites and that K_{IC} values comparable to those of 7075 Al(T6) are possible [59, 60]. In particular, McDanels [60] attributed the higher fracture strains observed in SiC$_w$/Al alloy composites compared to those reported in the literature for this type of composite to cleaner matrix powder, better mixing, and increased mechanical working during fabrication. McDanels' work also reinforces the idea that the metallic matrix is not merely a medium to hold the fibers together. Figure 6.20 shows that the type of aluminum matrix used in 20 v/o SiC$_w$/Al composites was the most important factor affecting yield strength and ultimate tensile strength of these composites. Higher-strength aluminum alloys showed higher strengths but lower ductilities. Composites with 6061 Al matrix showed good strength and higher fracture strain. The 5083 Al, in Fig. 6.20, is a nonheat treatable grade alloy.

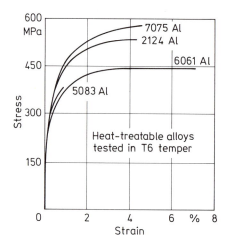

Fig. 6.20. Effect of aluminum matrix type on stress–strain behavior of 20 v/o SiC$_w$/Al composites. (From Ref. 60, used with permission.)

6.6 Applications

Among the general advantages of MMCs vis-à-vis PMCs, we can include higher-temperature capability, greater thermal and electrical conductivities, greater transverse tensile strength, greater shear and compressive strengths, and, last but not least, no problems of flammability and no significant hygral effects. The B/Al composites were one of the first MMCs to be developed. Boron fibers have high elastic modulus values and high tensile and compressive strengths coupled with a low density. B/Al composites are thus well suited for compression loading applications. The excellent compressive properties of B/Al are made use of in the tubular truss members in the Space Shuttle Orbiter.

Selective reinforcement of metals by ceramic fibers has become a fairly routine thing. Figure 6.21 shows an excellent example of partial reinforcement of pure aluminum. Sites marked A and B contain about 10% and 20% of SiC whiskers. Use of MMCs in automotive parts has increased since Toyota Motor Co. of Japan first introduced in the early 1980s selectively reinforced pistons in diesel engines [10]. This made other manufacturers follow suit. What is most impressive about these MMC pistons is that they cost about 15% less than conventional pistons.

Light, stiff, low expansion coefficient MMCs are very desirable in space applications. An interesting example is that of a large antenna boom cum waveguide designed for the NASA space telescope. In space, depending on orientation, parts of this boom may be under direct sunlight while other parts are shielded. This could mean a temperature difference of hundreds of degrees. If plain aluminum were to be used (aluminum has a high expansion coefficient of $\sim 24 \times 10^{-6} \mathrm{K}^{-1}$), it would distort excessively. A carbon fiber reinforced aluminum can be made such that $\alpha \simeq 0$ without any weight penalty over that of aluminum.

Figure 6.22 shows continuous carbon fiber reinforced aluminum and magnesium tubes. A Young's modulus greater than 385 GPa and a thermal expansion coefficient approaching zero have been claimed [61]. Extruded and machined internal combustion engine connecting rods are shown in Fig. 6.23a, while cast

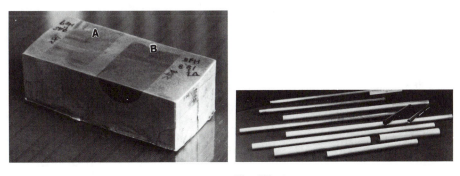

Fig. 6.21   **Fig. 6.22**

Fig. 6.21. Selective reinforcement of pure aluminum by SiC whiskers. Sites marked A and B contain about 10% and 20% whiskers. (Courtesy of Tokai Carbon Ltd.)

Fig. 6.22. Carbon fiber reinforced aluminum and magnesium tubes. (Courtesy of DWA Composite Specialties.)

Fig. 6.23. Some automotive parts made of MMCs: **a** extruded and machined components (courtesy of DWA Composite Specialties) **b** cast components (courtesy of Du Pont Co.)

Al_2O_3(FP)/Al and magnesium automotive parts are shown in Fig. 6.23b. These components show enhanced stiffness and wear resistance coupled with a higher use temperature. Among other applications of MMCs, Stöckel [62] has pointed out the use of tungsten fiber/copper matrix materials for electrical contact purposes, while NASA researchers [63] have examined the use of tungsten fiber/copper matrix, because of its excellent combination of strength, stiffness, and thermal and electrical conductivities, for use as cryogenically cooled thrust chamber liners for rocket engines.

A considerable body of work on higher use-temperature composites has been done at NASA [64]. The driving force for higher use-temperature materials is easy to understand. The efficiency of heat engines can be increased and fuel consumption can be reduced by obtaining a higher combustion temperature. Three types of composite were considered: (a) brittle/brittle (e.g., SiC or carbon in a ceramic or carbon matrix), (b) brittle/ductile (e.g., Al_2O_3 in iron-or nickel-based alloys), and (c) ductile/ductile (e.g., refractory metal alloy wire in iron or nickel-based alloys). The problem areas include reactions at the fiber/matrix interfaces resulting in fiber property degradation, thermal expansion mismatch between the fiber and matrix, and suitable processing methods. In spite of the fact that tungsten is a very heavy metal (density = 19.3 g cm^{-3}), the tungsten fiber reinforced superalloy (TFRS) composite system seems to be quite promising [64]. TFRS offers a use-temperature increase for turbine components of up to 150°C above that possible with commercial superalloys. Figure 6.24a shows the process for making the tungsten fiber reinforced superalloy blade. The fabrication process involves the diffusion bonding of monolayer composites with steel core plies and the use of unreinforced cover skin plies at the inner and outer surfaces. The steel core plies are leached out after diffusion bonding and a hollow airfoil is obtained. Figure 6.24b shows a hollow air-cooled blade made of TFRS as well as its microstructure. This TFRS blade has a weight approaching that of commercial superalloys.

Among the in situ or eutectic composites, aluminum, cobalt, nickel, or tantalum based eutectic composites result in a density between 5 and 8 g cm^{-3} and an ultimate tensile strength of about 1500 MPa. Such composites are potential candidates for high-temperature applications. One of the stronger eutectic alloy systems is the (Ni_3Al) strengthened γ(Ni) matrix; the reinforcing phase may be Ni_3Nb, Mo, TaC, NbC, or Cr_3C_2. Among the major problems are the slow freezing rates involved and the severe reactions with the foundry molds and cores [65]. An in situ composite system that has found commercial application is based on the physical

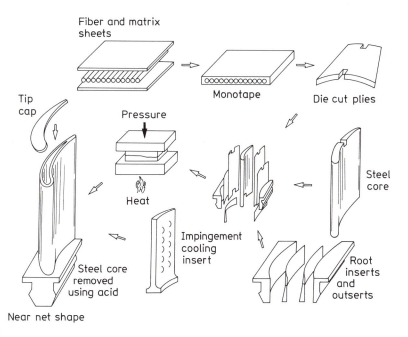

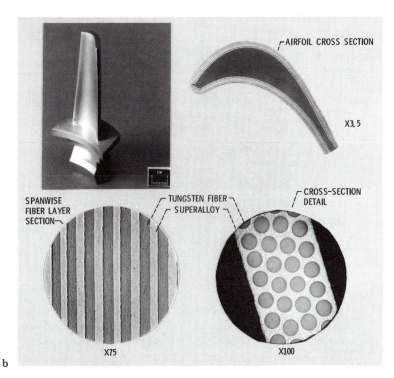

Fig. 6.24. a Steps in the tungsten fiber reinforced superalloy blade fabrication process. **b** A hollow air-cooled blade made of tungsten fiber reinforced superalloy and its microstructure. (Courtesy of NASA.)

Table 6.4. Metal matrix composite systems with metallic fibers

Fibers	Matrix	Fabrication method	Field of application
Stainless steel	Al	Powder metallurgy, lamination	Aircraft industry
Be ribbons	Al, Ti	Al or Ti clad Be rods are inserted in drilled Al or Ti performs; mechanical deformation of the preform	Shafts for high-speed rotating engines, e.g., rotor shafts in helicopters and control rods in aircraft
Ta	Mg	Infiltration technique	Aircraft industry
Mo	Ti or Ti alloy	Powder metallurgy, fiber alignment by extrusion, rolling, etc.	Supersonic aircraft rocket propulsion
Be or Ti clad Be	Ti alloy, pure Ti	Explosive welding of interposed layers of Ti and Be fibers	Aircraft construction
Be	Ti Ti-6 Al-4V Ti-6 Al-6V 2Sn Ti-5 Al-2 5Sn	Hot extrusion of mixed precursor of Ti and Be; the latter can be present as fiber preforms	Aircraft industry
Ni_3Al	Ni, (2–10%) Al	Powder metallurgy; the fiberlike phase is formed in situ during the mechanical working	Oxidation resistant, high strength at high and low temperatures
Ni-Cr-Al-Y	Ni alloy	Powder metallurgy	Sealing elements in turbines and compressors
W/1% ThO_2	Superalloy	Investment casting	Jet engines
W	W-Ni-Fe alloy	Liquid phase sintering; the W fibers are recrystallized to avoid dissolution	
Stainless steel	Ni alloys	Electroforming	Rocket engines
Mo, Ti, Nb	Ni superalloys	Powder metallurgy	
W	Cu	Melt impregnation	Electrical machinery
Nb filaments	Ni, Cu, Ag	Nb filaments embedded in a Cu, Ni, or Ag matrix are passed through a molten bath of Sn to form Nb_3Sn	Superconductors

Source: Adapted with permission from Ref. 67.

characteristics of the quasibinary system InSb-NiSb. A magnetoresistance device based on this system has been developed and is in use as a contactless control device that indicates current for the brake of an electrical locomotive [66].

An impressive amount of variety is available in MMCs. Naturally, some systems have been investigated more than others. We summarize the different MMCs containing metallic fibers, nonoxide ceramic fibers, and oxide ceramic fibers in Tables 6.4, 6.5, and 6.6, respectively [67]. Their methods of fabrication as well as the fields of application are also included.

Table 6.5. Metal Matrix composite with nonoxide ceramic fibers

Fibers	Matrix	Fabrication method	Field of application
SiC coated B	Al	Powder metallurgy; the composite article is clad with a sheet of Ti by diffusion bonding	Turbine blades
C (graphite, amorphous carbon)	Ni/Co aluminide	Coating C fibers with Ni or Co; mixing with Ni-Co-Al powder; hot pressing	
C coated with boride of Ti, Zr, Hf	Al or Al alloys, Mg, Pb, Sn, Cu, Zn	Melt impregnation	
C	Al alloy containing carbide-forming metal, e.g., Ti and Zr	Melt impregnation	
SiC with W core	Al-Cu alloy	Coating the filaments with Cu; passing the Cu-coated filaments through an Al melt	
C	Mg or Mg alloy	Melt impregnation; the molten Mg matrix contains small amounts of magnesium nitride to enhance wetting of the fibers	Turbine fan blades, pressure vessels, armor plates
SiC	Be or alloys with Ca, W, Mo, Fe, Co, Ni, Cr, Si, Cu, Mg, and Zr	Vacuum impregnation with molten Be or plasma spraying fibers with Be and consolidation by metallurgical process	Aerospace and nuclear industries
B + stainless steel; Borsic + Mo fibers	Al, Ti	Impregnation, spraying, etc.; combination of high-strength ductile and brittle fibers	Aerospace industry
SiC	Ti or alloy Ti-3 Al-2.5V	Hot pressing of interposed layers of fibers and matrix sheets; SiC fibers are previously coated with Zr diffusion barrier layer	Compressor blades, airfoil surfaces
Carbides of Nb, Ta, and W	Ni-Co and Fe-Cr alloys	Unidirectional solidification	Aircraft industry
SiC containing 0.01–20% free carbon	Cr-based alloys	Powder metallurgy; the free carbon reacts with the Cr to form carbides, thus improving bonding	High-strength, heat-resistant material, e.g., vanes and blades for turbines, rocket nozzles

(Continued)

Table 6.5 (*continued*)

Fibers	Matrix	Fabrication method	Field of application
SiC containing 0.01–30% free carbon	Co or Co-based alloys	Powder metallurgy or melt impregnation; carbide formation between the fibers and the Co matrix	High-strength, heat-resistant material, e.g., vanes and blades for turbines, rocket nozzles
SiC containing 0.01–20% free carbon	Mo-based alloys	Powder metallurgy	High-strength, heat-resistant material, e.g., vanes and blades for turbines, rocket nozzles
C coated with carbides	Ni or Ni alloys	Melt impregnation	Aeronautical industry
B	Cu-Ti-Sn alloy	Liquid phase sintering	Cutting tools
C	Bronze	Various processes	Bearing materials
C	Cu alloy	Powder metallurgy; the fibers are mixed with a slurry of Cu powder and 2% of a carbide-forming metal powder (Ti or Cr)	High strength, electrically conductive materials
C coated with Ti boride	Al, Cu, Sn, Pb, Ag, Zn, and Mg	The matrix contains alloying elements of Ti and B to prevent deterioration of the TiB coating of the fibers	Aeronautical industry
C coated with Ni	Metals with melting point lower than that of Ni	Melt impregnation	
C coated with SiO_2 + SiC	Al, Mg, Ti, and Ni	Melt impregnation, powder metallurgy	
Monocarbides of Ta, Ti, and W	Al, Al-Si alloy, Ag or Ag alloys, and Cu or Cu alloys	Melt impregnation	Abrasion-resistant materials
β-SiC	Ag or Ag alloys		Electrical conductors, contacts
C	Si	Powder metallurgy	Abrasive materials
SiC_w	Al		
C coated with TiB	Mg, Pb, Zn, Cu, Al	Melt impregnation	

Source: Adpated with permission from Ref. 67.

Table 6.6. Metal Matrix composite systems with oxide ceramic fibers

Fibers	Matrix	Fabrication method	Field of application
Al_2O_3, SiC, Al oxynitride	Al-Cu alloy	Mixing minute filaments in molten matrix; after solidification, the filaments penetrate through grain boundary regions	
Al_2O_3	Al-Li alloy	Infiltration with a molten 1–8% Li alloy; reaction occurs between Al_2O_3 fiber and the Li	
Al_2O_3-SiO_2, Y Al_2O_3	Al, Al-Zn alloy	Melt impregnation, powder metallurgy, etc.	Aeronautical industry
Ni-coated glass ceramic fibers	Al	Powder metallurgy	Sliding parts
Al_2O_3 continuous and polycrystalline	Mg or Mg alloy	Melt impregnation of aligned fibers	Turbine blades, shafts
C, B, glass, ceramic, and metal fibers	Mg alloy	Powder metallurgy; the composite contains two different Mg alloy phases	Multiple applications
Metal-coated glass fibers	Fe, Be, Ti, Al, and Sn	Chopped metal-coated glass fibers are mixed with glass filaments and metal powder; hot pressing	Dimensionally stable machine parts, e.g., friction elements
Glass fibers	Pb	Melt impregnation	Battery plates, bearing materials, acoustic insulation

Source: Adapted with permission from Ref. 67.

References

1   J.A. Cornie, Y-M. Chiang, D.R. Uhlmann, A.S. Mortensen, and J.M. Collins, *Ceram. Bull.*, **65**, 293 (1986).
2   R. Naslain, J. Thebault, and R. Pailler, in *Proceedings of the 1975 International Conference on Composite Materials*, vol.1, TMS-AIME, New York, 1976, p. 116.
3   M.F. Amateau, "Progress in the Development of Graphite Aluminum Composites by Liquid Infiltration Technology," Aerospace Corp. Rep. No. ATR-76 (8162)-3, 1976.
4   W. Meyerer, D. Kizer, S. Paprocki, and H. Paul, in *Proceedings of the 1978 International Conference on Composite Materials (ICCM/2)*, TMS-AIME,New York, 1978, p. 141.
5   K.G. Kreider and K.M. Prewo, in *Metal Matrix Composites* (vol. 4 in the series Composite Materials), Academic Press, New York, 1974, p. 400.
6   P.R. Smith and F.H. Froes, *J. Met.*, **36**, 19 (Mar. 1984).
7   P.R. Siemers, M.R. Jackson, R.L. Mehan, and J.R. Rairden, "Production of Composite Structures by Low Pressure Plasma Deposition", G.E. Report No. 85CRD001, Jan. 1985.
8   K.K. Chawla and C.E. Collares, in *Proceedings of the 1978 International Conference on Composite Materials (ICCM/2)*, TMS-AIME, New York, 1978, p. 1237.

9   P.M.B. Slate, in *Proceedings of the 1975 International Conference on Composite Materials*, vol. 1, TMS-AIME, Warrendale, PA, 1985, p. 743.

10 T. Donomoto, N. Miura, K. Funatani, and N. Miyake, "Ceramic Fiber Reinforced Piston for High Performance Diesel Engine," SAE Tech. Paper No. 83052, Detroit, MI, 1983.

11 M.W. Toaz and M.D. Smalc, *Diesel Prog. N. Am.*, (June 1985).

12 J.L. Walter, in *In Situ Composites IV*, Elsevier, New York, 1982, p. 85.

13 R.G. Hill, R.P. Nelson, and C.L. Hellerich, in *Proceedings of the 16th Refractory Working Group Meeting*, Seattle, WA, Oct. 1969.

14 K.K. Chawla and M. Metzger, in *Advances in Research on Strength and Fracture of Materials*, vol. 3, Pergamon Press, New York, 1978, p. 1039.

15 C. Manning and T. Gurganus, *J. Am. Ceram. Soc.*, **52**, 115 (1969).

16 S. Rhee, *J. Am. Ceram. Soc.*, **53**, 386 (1970).

17 A.A. Baker and C. Shipman, Fiber Sci. Tech., **5**, 285 (1972).

18 H.A. Katzman, J. Mater. Sci., **22**, 144 (1987).

19 C.G. Levi, G.J. Abbaschian, and R. Mehrabian, *Met. Trans.A*, **9A**, 697 (1978).

20 G.R. Cappelman, J.F. Watts, T.W. Clyne, *J. Mater. Sci.*, **20**, 2159 (1985).

21 A.R. Champion, W.H. Krueger, H.S. Hartman, and A.K. Dhingra, in *Proceedings of the 1978 International Conference on Composite Materials (ICCM/2)*, TMS-AIME, New YOrk, 1978, p. 883.

22 A.K. Dhingra, *Proc. R. Soc. London*, **A294**, 559 (1980).

23 K.M. Prewo, United Technologies Research Center, Rep. R77-912245-3, May 1977, as cited in Ref. 20.

24 I.W. Hall and V. Barrailler, *Met. Trans. A*, **17A**, 1075 (1986).

25 R.T. Pepper, J. W. Upp, R.C. Rossi, and E.G. Kendall, *Met. Trans.*, **2**, 117 (1971).

26 Y. Kimura, Y. Mishima, S. Umekawa, and T. Suzuki, *J. Mater. Sci.*, **19**, 3107 (1984).

27 R. Warren and C-H. Andersson, *Composites*, **15**, 101 (1984).

28 S. Kohara, in *Proceedings of the Japan–United States Conference on Composite Materials*, Japan Society for Composite Materials, Tokyo, 1981.

29 H.E. Cline, J. L. Walter, E.F. Koch, and L.M. Osika, *Acta Met.*, **19**, 405 (1971).

30 J. van Suchtelen, Philips Res. Rep., **27**, 28 (1972).

31 L.-J. Fu, M. Schmerling, and H.L. Marcus, in *Composite Materials: Fatigue and Fracture*, ASTM STP 907, American Society for Testing and Materials, Philadelphia, 1986, p.51.

32 J.A. DiCarlo, *J. Met.* **37**, 44 (June 1985).

33 K.K. Chawla and A.C. Bastos, in *Proceedings of the 1975 International Conference on Composite Materials*, vol. 1, TMS-AIME, New York, 1976, p. 549.

34 K.K. Chawla and M. Metzger, *J. Mater. Sci.*, **7**, 34 (1972).

35 K.K. Chawla, *Philos Mag.*, **28**, 401 (1973).

36 K.K. Chawla, *Metallography*, **6**, 55 (1973).

37 K.K. Chawla, in *Grain Boundaries in Engineering Materials, Proceedings of the 4th Bolton Landing Conference*, Claitor's Publishing, Baton Rouge, LA, 1974, p. 435.

38 K.K. Chawla, A.C. Bastos, and F.A. Cunha Silva, *Trans. Japan Soc. Composite Mater.* **3**, 14 (1977).

39 R.J. Arsenault and R.M. Fisher, *Scripta Met.*, **17**, 67 (1983).

40 M. Vogelsang, R.J. Arsenault, and R.M. Fisher, *Met. Trans. A*, **17A**, 379 (1986).

41 K.K. Chawla, in *Microstructural Science*, vol. 2, Elsevier, New York, 1974, p. 115.

42 W.G. Patterson and M. Taya, in *Proceedings of the Fifth International Conference on Composite Materials (ICCM/V)*, TMS-AIME, Warrendale, PA, 1986, p. 53.

43 T.G. Nieh and R.F. Karlak, *Scripta Met.*, **18**, 25 (1984).

44 O.B. Pedersen, in *Proceedings of the Fifth International Conference on Composite Materials (ICCM/V)*, TMS-AIME, Warrendale, PA., 1986, p. 1.

45 K.K. Chawla, *J. of Metals*, **37**, 25 (Dec. 1985).

46 A. Mortensen, M.N. Gugnor, J.A. Cornie, and M.C. Flemings, *J. Met.*, **38**, 30 (Mar. 1986).

47 A. Kohyama, N. Igata, Y. Imai, H. Teranishi, and T. Ishikawa, in *Proceedings of the Fifth International Conference on Composite Materials (ICCM/V)*, TMS-AIME, Warrendale, PA, 1985, p. 609.

48 W.L. Phillips, in *Proceedings of the 1978 International Conference on Composite Materials (ICCM/2)*, TMS-AIME, New York, 1978, p. 567.

49 E.G. Kendall, in *Metallic Matrix Composites* (vol. 4 in the series Composite Materials), Academic Press, New York, 1974, p. 319.

50   A. Kelly and K.N. Street, *Proc. R. Soc. London*, **328A**, 283 (1972).
51   H. Lilholt, in *Fatigue and Creep of Composite Materials*, Third Risø International Symposium, Risø, Denmark, 1982.
52   M. McLean, in *Proceedings of the Fifth International Conference on Composite Materials (ICCM/V)*, TMS-AIME, Warrendale, PA, 1986, p.37.
53   D.R. Williams and M.E. Fine, in *Proceedings of the Fifth International Conference on Composite Materials (ICCM/V)*, TMS-AIME, Warrendale, PA., 1985, p. 639.
54   K.K. Chawla, *Fiber Sci. Tech.*, **7**, 49 (1975).
55   G. Rosenkranz, V. Gerold, D. Stockel, and L. Tillmann, *J. Mater. Sci.*, **17**, 264 (1982).
56   G. Rosenkranz, V. Gerold, K. Kromp, D. Stockel, and L. Tillmann, *J. Mater. Sci.*, **17**, 277 (1982).
57   W.A. Logsdon and P.K. Liaw, *Eng. Fract. Mech.*, **24**, 737 (1986).
58   C.R. Crowe, R.A. Gray, and D.F. Hasson, in *Proceedings of the Fifth International Conference on Composite Materials (ICCM/V)*, TMS-AIME, Warrendale, PA, 1986, p. 843.
59   *Aviation Week and Space Technology*, **123**, 127 (Nov. 18, 1985).
60   D.L. McDanels, *Met. Trans.A*, **16A**, 1105 (1985).
61   W.C. Harrigan, DWA Composite Specialties, personal communication.
62   D. Stöckel, in *Proceedings of the 1975 International Conference on Composite Materials*, vol. 2, TMS-AIME, New York, 1976, p. 484.
63   D.L. McDanels. T.T. Serafini, and J.A. DiCarlo, NASA Tech. Memorandum 87132, National Aeronautics & Space Administration Washington, DC, 1985.
64   D.W. Petrasek and R.A. Signorelli, "Tungsten Fiber Reinforced Superalloys — A Status Review," NSAS Tech. Memorandum 82590, National Aeronautics & Space Administration, Washington, DC, 1981.
65   F.D. Lemkey, in *Industrial Materials Science & Engineering*, Marcel Dekker, New York, 1984, p. 441.
66   H.Weiss, *Met. Trans.*, **2**,1513 (1971).
67   P. Bracke, H. Schurmans, and J. Verhoest, *Inorganic Fibres and Composite Materials*, Pergamon Press, Oxford, 1983, p.89.

Suggested Reading

A.Banerji, P.K. Rohatgi, and W. Reif, *Metalwiss. Tech.*, **38**, 656 (1984).
T.W. Chou, A. Kelly, and A. Okura, *Composites*, **16**, 187 (July 1985).
K.G. Kreider (ed.), *Metal Matrix Composites* (vol. 4 in the series Composite Materials), Academic Press, New York, 1974.
C.T. Lynch and J.P. Kershaw, *Metal Matrix Composites*, CRC Press, Cleveland, OH, 1972.
M. McLean, *Directionally Solidified Materials for High Temperature Service*, The Metals Society, London, 1983.
S.T. Mileiko, in *Fabrication of Composites*, North-Holland, Amsterdam, 1983, p. 221.
Proceedings of the Conference on In Situ Composites, National Materials Advisory Board, NMAB-308, Washington, DC, 1973.
Proceedings of the Second Conference on In Situ Composites, Xerox Individualized Publishing, Lexington, MA, 1976.
Proceedings of the Third Conference on In Situ Composites, Ginn Custom Publishing, Lexington, MA, 1979.
Proceedings of the Fourth Conference on In Situ Composites, Elsevier, New York, 1982.
P.K. Rohatgi, R. Asthana, and S. Das, Solidification, Structures, and Properties of Cast Metal-ceramic Particle Composites, *Intl. Met. Rev.*, **31**, 115 (1986).

7. Ceramic Matrix Composites

Ceramic materials in general have a very attractive package of properties: high strength and high stiffness at very high temperatures, chemical inertness, low density, and so on. This attractive package is marred by one deadly flaw, namely, an utter lack of toughness. They are prone to catastrophic failures in the presence of flaws (surface or internal). They are extremely susceptible to thermal shock and any damage done to them during fabrication and/or service. It is therefore understandable that on overriding consideration in ceramic matrix composites (CMCs) has been to toughen the ceramic matrices by incorporating fibers in them and thus exploit the attractive high-temperature strength and environmental resistance of ceramic materials without risking a catastrophic failure.

It is worth pointing out at the very outset that there are certain basic differences between CMCs and other composites. The general philosophy in nonceramic matrix composites is to have the fiber bear a greater proportion of the applied load. This load partitioning depends on the ratio of fiber and matrix elastic moduli, E_f/E_m. In nonceramic matrix composites, this ratio can be very high, while in CMCs, it is rather low and can be as low as unity. Another distinctive point regarding CMCs is that because of limited matrix ductility and generally high fabrication temperatures, thermal mismatch between components has a very important bearing on CMC performance. The problem of chemical compatibility between components in CMCs has ramifications similar to those in, say, MMCs. We describe first some of the fabrication methods for CMCs, followed by a description of some salient characteristics of CMCs regarding interface and mechanical properties and, in particular, the various possible toughness mechanisms.

7.1 Fabrication of CMCs

Phillips [1] and Cornie et al. [2] have reviewed the CMC manufacturing techniques. Generally, a two-stage process is used for CMCs: incorporation of a reinforcing phase into an unconsolidated matrix followed by matrix consolidation. The fiber incorporation stage also involves some kind of fiber alignment. By far, the most common technique of fiber incorporation is the slurry infiltration process. A fiber tow in passed through a slurry tank (containing the matrix powder, a carrier liquid, and an organic binder) and wound on a drum and dried. This is followed by cutting, stacking the tows, and consolidation (Fig. 7.1). The fibers should suffer as little damage as possible, while the matrix should have as little porosity as possible.

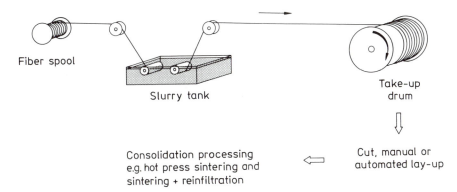

Fiber spool

Slurry tank

Take-up
drum

Consolidation processing   ⇐   Cut, manual or
e.g. hot press sintering and   automated lay-up
sintering + reinfiltration

Fig. 7.1. Schematic of the slurry infiltration process. (Adapted from Ref. 2, used with permission.)

Porosity in a ceramic material is a common and serious flaw. To this end, complete removal of the fugitive binder and a matrix powder particle smaller than the fiber diameter are important [1].

Hot pressing is the most common and perhaps one of the most expensive techniques used to consolidate CMCs. The reason for this is that it is very difficult to process ceramics any other way. The CMCs produced by hot pressing, however, are of a very superior quality, provided the thermal mismatch between the components is low. Carbon, alumina, silicon carbide, and metallic fibers have been incorporated in glass, glass-ceramic, and oxide-ceramic matrices. Powder size distribution, temperature, and pressure are the critical parameters that control the matrix porosity, fiber damage, and fiber/matrix bond strength. For example, Fig. 7.2 shows the effect of hot-pressing temperature and pressure on the bond strength of carbon fiber reinforced borosilicate glass [3]. Samples produced at 1000°C had lower flexural strength because they had a higher porosity than those produced at higher temperatures. Samples subjected to 6.3 MPa pressure were stronger than those subjected to 10.5 MPa because the higher pressure resulted in a greater damage to the fibers. Other composite characteristics that can be affected by these processing parameters include fiber/matrix bonding and reactions at the interface. We discuss these in Sect. 7.3.

A general limitation of processes involving consolidation of impregnated tows is that only one- or two-dimensional reinforcement geometries can be produced easily. Matrix-rich regions are also quite common. Cold pressing of fiber and matrix powder followed by sintering is another technique. Generally, in this technique the matrix shrinks a lot during sintering and the resultant composite has a lot of cracks.

Sambell et al. [3] produced carbon fiber reinforced glass-ceramics by first hot pressing carbon fibers and glass at low temperatures, followed by a high-temperature treatment to crystallize the glassy matrix. Prewo and coworkers [4–6] studied a variety of fiber reinforced glass and glass-ceramic matrix composites made with the slurry infiltration and mixing techniques and obtained impressive results (see Sect. 7.2).

Melt infiltration, in situ chemical reaction, and sol-gel and polymer pyrolysis are the other techniques that have been used to produce CMCs. Melt processing (Fig. 7.3) yields a high-density and virtually pore-free matrix, and practically any rein-

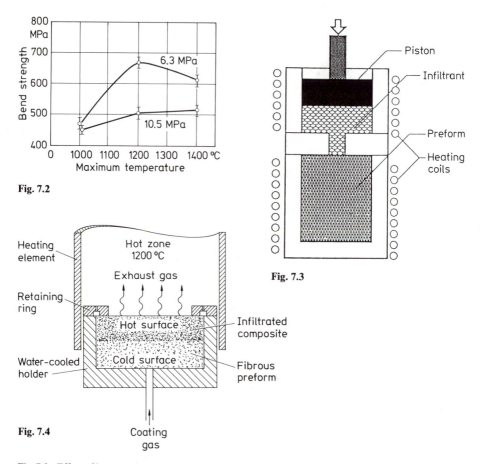

Fig. 7.2

Fig. 7.3

Fig. 7.4

Fig. 7.2. Effect of hot-pressing temperature and pressure on the bend strength of carbon fiber reinforced borosilicate glass. (Adapted from Ref. 3, used with permission.)

Fig. 7.3. Schematic of melt infiltration process. (Reprinted from Ref. 2, used with permission.)

Fig. 7.4. Schematic of chemical vapor infiltration process. (Reprinted from Ref. 10, used with permission.)

forcement geometry can be used. The temperatures involved, however, are very high. In situ chemical reaction techniques to produce CMCs are extensions of techniques to produce monolithic ceramic bodies. These include chemical vapor deposition (CVD), chemical vapor infiltration (CVI), and reaction bonding. The CVI method has been successfully employed to impregnate fibrous preforms [7–10]. Commonly, the process involves thermal decomposition of methyltrichlorosilane (CH_3SiCl_3) to SiC between 1200 and 1400 K. The vapors of SiC (or for that matter those of any other suitable ceramic matrix material) deposit as solid phases on and between the fibers to form the matrix. The CVI process of matrix formation can be characterized as a low-stress and low-temperature CVD process that avoids some of the problems associated with high-temperature ceramic processing. A schematic of the process used by Stinton et al. [10] is shown in Fig. 7.4. This process combines thermal gradient and pressure gradient approaches. A graphite holder in contact with a water-cooled metallic gas distributor holds the fibrous preform. The bottom and side surfaces thus stay cool while the top of the fibrous preform is exposed to the hot zone, creating a steep thermal gradient. The reactant gaseous mixture passes unreacted through the fibrous preform because of the low temperature. When these gases reach the hot zone, they decompose and deposit on and between the fibers to form the matrix. As the matrix material gets deposited in the hot portion of the preform, the preform density and thermal conductivity increase and the hot zone moves progressively from the top of the preform toward the bottom. When the composite is formed completely at the top and is no longer permeable, the gases flow radially through the preform, exiting from the vented retaining ring. The final obtainable density in a ceramic body is limited by the fact that closed porosity starts at about 93–94% of theoretical density. It is difficult to impregnate past this point.

Sol-gel and polymer pyrolysis techniques have also been used to make ceramic matrix materials in a fibrous preform (cf. sol-gel preparation of ceramic fibers in Sect. 2.1). The advantages include lower processing temperatures, greater compositional homogeneity in single-phase matrices, a potential for producing unique multiphase matrix materials, and so on. Covalent ceramics, for example, can be produced by pyrolysis of polymeric precursors at temperatures as low as 1400°C and with yields greater than those in CVD processes [2]. Among the disadvantages are high shrinkage and low yield compared to slurry techniques. Frequently, repeated impregnations are required to produce a substantially dense matrix. Fitzer and Gadow [9], for example, give the following recipe for making fiber reinforced organosilicon-based SiC composites:

1. Prepare the porous fibrous preform with some binder phase.
2. Evacuate in an autoclave.
3. Infiltrate the samples with silazanes or polycarbosilanes at high temperature and high pressure and polymerize.
4. Remove and fix the infiltrated samples.
5. Thermally decompose the organosilicon polymer matrix in an inert atmosphere (autoclave) between 800 and 1300 K.
6. Repeat steps 2–5 to attain a high density.
7. Anneal at 1300–1800 K to produce an optimum crystal structure.

The reader is referred to Ref. 9 for details.

7.2 Properties of CMCs

As mentioned above, the relative elastic modulus values of fiber and matrix are very important in CMCs. The ratio E_f/E_m determines the extent of matrix microcracking. The ceramic matrix fracture strains tend to be rather low. In MMCs and in thermoplastic PMCs the matrix failure strain (e_f) is considerably greater than that of fibers. Most unreinforced metals show $e_f > 10\%$ while most polymers fail between 3 and 5% strain. Thus, in both MMCs and PMCs fiber failure strain controls the composite failure strain. Typically, fibers such as boron, carbon, and silicon

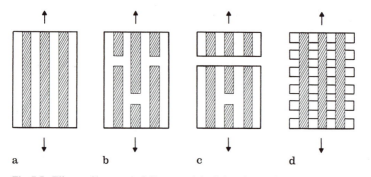

Fig. 7.5. Fiber and/or matrix failure: **a** original situation, **b** fiber failure, **c** composite failure controlled by fiber failure, **d** fibers bridging the matrix cracks

carbide show failure strain values of $\sim 1\%$. Compare this with the failure strains of less than 0.05% for most ceramic matrix materials. The situation in regard to fiber and/or matrix failure is shown in a simplified manner in Fig. 7.5. In the case of MMCs and PMCs, fibers fail first at various weak points distributed along their lengths; see Fig. 7.5b. The composite will fail along a section that has a maximum number of fiber fractures; see Fig. 7.5c. In a strongly bonded CMC, fiber and matrix would fail simultaneously at matrix failure strain and a situation similar to that shown in Fig. 7.5c will prevail. In a weakly bonded CMC, however, the matrix will start cracking first and the fibers will be bridging the matrix blocks (Fig. 7.5d). Thus, from a toughness point of view, we do not desire too strong a bond in a CMC because it would make a crack run through the specimen. A weaker interface, however, would lead to fiber-bridging of matrix microcracks. Let us consider the simple isostrain model that predicts a rule of mixtures type relationship (see Chap. 10) for a unidirectional composite. Let the composite be subjected to a strain e_c. The isostrain condition implies

$$e_c = e_f = e_m$$

where subscripts c, f, and m denote composite, fiber, and matrix, respectively. This results in a rule of mixtures relationship for strength in the longitudinal direction (see Chap. 10), namely,

$$\sigma_c = \sigma_f V_f + \sigma_m V_m$$

where σ is the stress, V is the volume fraction, and the subscripts have the same meaning as given above. Note that $V_f + V_m = 1$. Then, assuming elastic behavior for both the matrix and fiber, we can write

$$\frac{\text{stress carried by fiber}}{\text{stress carried by matrix}} = \frac{\sigma_f V_f}{\sigma_m V_m} = \frac{E_f V_f}{E_m V_m}$$

As the composite is loaded, matrix failure strain being smaller than that of the fiber, it will start showing microcracks at some stress σ_0, as shown in Fig. 7.6. We can write

$$\sigma_0 = \sigma_f V_f + \sigma_{mu}(1 - V_f)$$

where σ_{mu} is the matrix stress at its breaking strain.

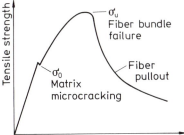

Fig. 7.6. Tensile stress–strain curve (schematic) of an aligned CMC in the longitudinal direction. (Reprinted from Ref. 11, used with permission.)

Table 7.1. Theoretical matrix cracking stresses for a borosilicate glass and magnesia reinforced with 60% of high-modulus carbon fibers (E = 360 GPa)

	E_m (GPa)	σ_{mu} (MPa)	σ_0 (MPa)
Borosilicate glass	60	100	400
Magnesia	250	120	151

Source: Adapted with permission from Ref. 1.

Rearranging, we have

$$\sigma_0 = \sigma_{mu}\left(\frac{\sigma_f V_f}{\sigma_{mu}} + (1 - V_f)\right) = \sigma_{mu}\left[1 + V_f\left(\frac{E_f}{E_m} - 1\right)\right]$$

Table 7.1 shows some relevant parameters for borosilicate glass and magnesia matrix materials containing 60% V_f of carbon fibers [1]. We note that in high matrix modulus composites (magnesia in the present case), matrix cracking would occur at much lower stresses. Thus, it would appear that low-modulus glasses and ceramics offer some advantages over high-modulus ceramic matrices. A CMC with even a microcracked ceramic matrix can retain some reasonable strength ($\sigma_c \simeq \sigma_f V_f$) and there are applications, for example, bushings, where such a damage-tolerant characteristic is very valuable since, in the absence of fibers bridging the cracks, the monolithic matrix would disintegrate. The disadvantage of course is that matrix microcracking provides an easy path for environmental attack of the fibers and the fiber/matrix interface.

Let us focus attention on the tensile stress–strain curve of an aligned CMC in the longitudinal direction, as shown schematically in Fig. 7.6 [11]. This figure shows that CMCs have damage-tolerant characteristics in uniaxial tension. At a stress σ_0, the stress–strain curve shows a dip, indicating the incidence of periodic matrix cracking. Since the fibers have enough strength to support the load in the presence of a damaged matrix (a very desirable feature indeed), the stress–strain curve continues to rise until, at a stress marked σ_u, the fiber bundle fails. At this point, the phenomenon of fiber pullout starts. The extent of this fiber pullout region depends critically on the interfacial frictional resistance, τ_i. The fiber/matrix interface has a lot to do with the form of the stress–strain curve. If the bonding is too strong, matrix

Table 7.2. Properties of carbon fiber reinforced glass composite

	Bend strength (MPa)	Young's modulus (GPa)	Work of fracture (kJ m^{-2})
Unreinforced glass (Pyrex)	100	60	0.004
Carbon/glass (Pyrex) 50% V_f, unidirectional	700	193	5.0

Source: Adapted with permission from Ref. 12.

cracking will be accompanied by a small amount of fiber pullout, which is an undesirable characteristic from the toughness viewpoint as we shall see in Sect. 7.4. Both σ_0 and σ_u are insensitive to specimen or component size because both strength levels are independent of matrix flaws [11]. This is in distinct contrast to the behavior of monolithic ceramic materials which show a significant size dependence. Increased stiffness and strength were observed in a unidirectionally aligned composite consisting of continuous carbon fibers (50% V_f) in a glass matrix compared to the unreinforced glass matrix [12]. But more importantly, there occurred a large increase in the work of fracture. Table 7.2 summarizes these properties. The increased work of fracture is a result of the controlled fracture behavior of the composite, while the unreinforced matrix failed in a catastrophic manner.

Fiber length, or more precisely, the fiber aspect ratio (length/diameter), fiber orientation, relative strengths and moduli of fiber and matrix, thermal expansion mismatch, matrix porosity, and fiber flaws are the important variables that control the performance of CMCs. Sambell et al. [3] showed that, for ceramic matrix materials containing short, randomly distributed carbon fibers, there occurred a weakening effect rather than a strengthening effect. This was attributed to the stress concentration effect at the extremities of randomly distributed short fibers and thermal expansion mismatch.

Aligned continuous fibers do lead to a real fiber reinforcement effect. The stress concentration at fiber ends is minimized and higher fiber volume fractions can be obtained. At very high fiber volume fractions, however, it becomes difficult to remove matrix porosity. Figure 7.7 shows a linear increase in strength with fiber volume fraction V_f up to $\sim 55\%$ [13]. Beyond 55% V_f, the matrix porosity increased. The Young's modulus also increased linearly with V_f (see Fig. 7.8), but at higher V_f it deviated from linearity owing to matrix porosity and possible fiber misalignment [13]. Impressive work on carbon and SiC fiber reinforced glass and glass-ceramic composites has been done by Prewo and coworkers [5–7, 14]. Their work showed that tough and strong CMCs could be made. Extensive fiber pullout and a controlled fracture behavior of the CMC, that is, a slow loss in load-bearing capacity with increasing strain, were the marked characteristics. The impressive properties obtained in SiC/LAS (lithium aluminosilicate) glass-ceramic matrix are summarized in Table 7.3 [14]. Note the high-temperature strength and fracture toughness levels obtained. Nonoxide ceramic matrix composites have also been investigated. Their

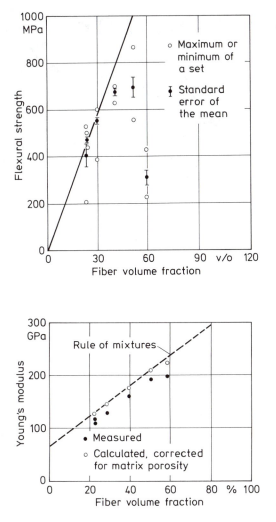

Fig. 7.7. Flexural strength as a function of fiber volume fraction: continuous, aligned carbon fibers in a borosilicate glass matrix. (Reprinted from Ref. 13, used with permission.)

Fig. 7.8. Young's modulus as a function of fiber volume fraction: continuous, aligned carbon fibers in a borosilicate glass matrix. (Reprinted from Ref. 13, used with permission.)

Table 7.3. Properties of 50% SiC/LAS

Property	Fiber orientation	
	0°	0°–90°
Density (g cm^{-3})	2.5	2.5
Flexural strength (MPa)		
Room temperature	600	380
800°C	800	410
1000°C	850	480
Fracture toughness (MPa m$^{1/2}$)		
Room temperature	17	10
1000°C	25	12

Source: Adapted with permission from Ref. 14.

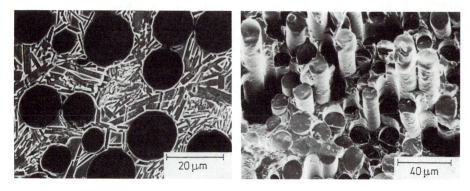

Fig. 7.9   Fig. 7.10

Fig. 7.9. Microstructure of hot-pressed SiC(Nicalon)/Ba-Si-Al-O-N composite. Note the crystalline structure of the matrix. (Reprinted from Ref. 15, used with permission.)

Fig. 7.10. Fracture surface of SiC(Nicalon)/Ba-Si-Al-O-N composite showing fiber pullout. (Reprinted from Ref. 15, used with permission.)

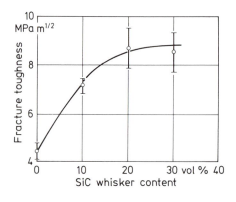

Fig. 7.11. Toughness gains obtained by incorporating SiC_w in alumina (Reprinted from Ref. 18, used with permission.)

main attraction is their refractoriness and possible superior mechanical properties. Herron and Risbud [15] studied a Ba-Si-Al-O-N glass ceramic containing Nicalon SiC fibers. A scanning electron micrograph of the hot-pressed composite is shown in Fig. 7.9. The fracture surface of this composite as observed in a scanning electron microscope is shown in Fig. 7.10, where one sees the phenomenon of fiber pullout, indicating a weak fiber/matrix bond. The fibers were bonded to the matrix by an amorphous layer whose characteristics changed with heat treatment; a carbon-rich layer was also observed on the fiber surface (15). Remnants of the interfacial amorphous layer adhering to the SiC fibers can be seen in Fig. 7.10.

Among oxide ceramic matrix materials, alumina and zirconia have attracted the most attention. In particular, SiC whisker reinforced alumina composites show a good deal of promise. Wei, Becher, and Tiegs at Oak Ridge National Laboratory [16–18] have been able to obtain impressive gains in toughness and strength by incorporating 20–30% by volume of SiC whiskers in alumina by hot pressing. Figures 7.11 and 7.12 show these gains. A typical fine-grained monolithic alumina has a toughness (K_{Ic}) of 4–5 MPa m$^{1/2}$ and a flexural strength between 350 and

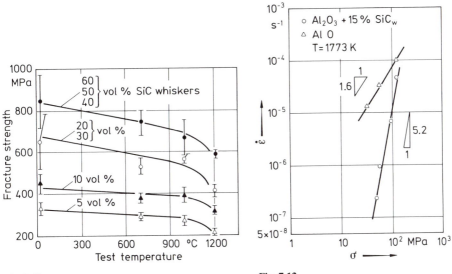

Fig. 7.12   Fig. 7.13

Fig. 7.12. High-temperature strength increase as a function of the SiC_w volume fraction in SiC_w/Al_2O_3. (Reprinted from Ref. 18, used with permission.)

Fig. 7.13. Increased creep resistance of the composite (SiC_w/Al_2O_3) vis-à-vis Al_2O_3. (Reprinted from Ref. 19, used with permission.)

Fig. 7.14. Dislocations emanating from the SiC whisker tip: TEM. (Reprinted from Ref. 19, used with permission.)

450 MPa. Al_2O_3 containing 20% by volume of SiC whiskers has a K_{Ic} of 8–8.5 MPa $m^{1/2}$ and a flexural strength of 650 MPa; these levels are maintained up to about 1000°C. Chokshi and Porter [19] found SiC whisker reinforced alumina to be more creep resistant than polycrystalline alumina; see Fig. 7.13. They also observed that

the composite showed a higher stress exponent for creep (see Fig. 7.13). The higher stress exponent indicated a change in the operating creep mechanism. The exponent for polycrystalline alumina is about 2 and this is rationalized in terms of some kind of diffusion creep being the controlling mechanism. A stress exponent of about 5, which was observed for the composite, is indicative of a dislocation creep mechanism being in operation. Observation of the specimens deformed in creep in transmission electron microscope showed dislocation activity. Figure 7.14 shows dislocations emanating from the whisker tips, probably resulting from high stress concentrations at these sites.

7.3 Interface in CMCs

In general, for CMCs one must satisfy the following compatibility requirements: (a) thermal expansion compatibility and (b) chemical compatibility.

Ceramics have a limited ductility and in the fabrication of CMCs one uses high fabrication temperatures. Thus, thermal mismatch on cooling can cause matrix (or fiber) cracking. Thermal strain in composites is proportional to $\Delta\alpha\,\Delta T$, where $\Delta\alpha = \alpha_f - \alpha_m$, α_f and α_m being the linear expansion coefficients of fiber and matrix respectively, and ΔT is the temperature interval. There is of course another complication, namely, fiber expansion coefficients are also sometimes not equal in the axial and radial directions. Carbon fiber in particular has the following axial and radial coefficients:

$$\alpha_a \approx 0$$

$$\alpha_r \approx 8 \times 10^{-6} \text{ K}^{-1}$$

If $\Delta\alpha_a$ is positive, the matrix is compressed on cooling, which is beneficial because it leads to an increase in the tensile stress at which matrix cracking will occur. Conversely, if $\Delta\alpha_a$ is negative, the matrix experiences tension, which, if ΔT is sufficiently large, can cause matrix cracking. In the radial direction, if $\Delta\alpha_r$ is positive, the fibers tend to shrink away from the matrix on cooling, which results in a reduced interfacial bond strength. If, however, $\Delta\alpha_r$ is negative, the fiber matrix bond strength can even be improved.

Matrix cracking resulting from thermal mismatch is a more serious problem in short-fiber composites than in continuous-fiber composites. The reason is that in a ceramic matrix containing aligned continuous fibers, transverse microcracks appear in the matrix, but fibers continue to hold the various matrix blocks together and the composite can still display a reasonable amount of strength (Fig. 7.5d). In a randomly oriented short-fiber composite, owing to increased stress at the fiber ends, matrix cracking occurs in all directions and the composite is very weak. We can define a thermal expansion mismatch parameter for axial and radial directions in an aligned fiber composite as [20]

$$\phi_a, \phi_r = \Delta\alpha\,\Delta T\left(\frac{E_m}{\sigma_m}\right)$$

Table 7.4. A comparison of damage resulting from the thermal expansion mismatch in some carbon fiber reinforced systems[a]

Matrix	$\alpha_m{}^b$ $(10^{-6}{}^\circ C^{-1})$	T_c (°C)	E (GPa)	$\sigma_{mu}{}^d$ (MPa)	$\phi_a{}^e$	$\phi_r{}^e$	Damage
MgO	13.6	1200	300	200	25	10	Severe cracking
Al$_2$O$_3$ (80% dense)	8.3	1400	230	300	9	0.3	Severe cracking
Soda-lime glass	8.9	480	60	100	2.6	0.3	Localized cracks
Borosilicate							
glass	3.5	520	60	100	1.1	−1.4	Uncracked
Glass-ceramic	1.5	1000	100	100	1.5	−6.5	Uncracked

[a] Type I carbon fibers $\alpha_a \approx 0$, $\alpha_r \approx 8 \times 10^{-6}$ K^{-1}.
[b] α_m is the matrix thermal expansion coefficient.
[c] T_c is the temperature below which little stress relaxation can occur.
[d] σ_{mu} is the matrix strength.
[e] ϕ_a and ϕ_r are the thermal expansion mismatch parameters.
Source: Adapted with permission from Ref. 1.

Table 7.4 makes a comparison of damage resulting from thermal expansion mismatch in some carbon fiber reinforced ceramic matrix composites. Note that only glass and glass-ceramic matrices show no damage.

Chemical compatibility between the ceramic matrix and the fiber involves the same thermodynamic and kinetic considerations as with other composite types. Quite frequently, the bond between fiber and ceramic matrix is simple mechanical interlocking. During fabrication (by hot pressing) or during subsequent heat treatments, the fiber–matrix bond could be affected by the high temperatures attained because of any chemical reaction between the fiber and matrix or because of any phase changes in either one of the components. Sambell et al. [20] studied the zirconia reinforced magnesia composite system in which there occurs a chemical reaction at 1600°C. At temperatures less than 1600°C, the composites showed a weak fiber/matrix interface and fiber pullout occurred during mechanical polishing. Upon heat treating at 1600°C, however, because of the interfacial reaction and the resultant improved bonding, no damage was observed upon mechanical polishing. Heat treatment at 1700°C resulted in the complete destruction of the zirconia fibers and the distribution of zirconia to grain boundaries in magnesia. Thus, as we noted in the case of MMCs (Chap. 6), it is of the utmost importance to be able to control the interfacial bond by means of controlled chemical reaction between components. Chokshi and Porter [19] also observed a reaction layer on the fiber surface after creep testing in air. Auger electron microscopy analysis showed a mullite layer with large glassy phase regions along grain boundaries. The following interfacial reaction was proposed:

$$2\,SiC + 3\,Al_2O_3 + 4\,O = Al_6Si_2O_{13} + 2\,C$$

and reaction kinetics were modeled by an equation of the form

$$x^2 = Dt$$

where x is the reaction zone thickness, $D = D_0 \exp(-Q/kT)$, Q is the activation energy, k is Boltzmann's constant, T is the temperature in kelvin, and t is the time in seconds (see Sect. 6.3).

The nature of the bond between fiber and ceramic matrix is thought to be predominantly mechanical. Most researchers believe that in carbon fiber reinforced glass or glass-ceramics there is little or no chemical bonding [12–14]. Evidence for this is the low transverse strength of these composites and the fact that one does not see matrix material adhering to the fibers on the fracture surface. The bond is thought to be entirely mechanical with the ceramic matrix penetrating the irregularities present on the carbon surface. Shear strength data on carbon fiber in borosilicate glass and lithium-aluminosilicate glass-ceramics (LAS) show that the borosilicate glass composites have double the shear strength of LAS composites. The reason for this is the different radial shrinkage of fibers from the matrix during cooling. Calculated radial contractions of fiber from the matrix are 2.4×10^{-8} m for the LAS ceramic and 0.9×10^{-8} m for the glass ceramic composite [1]. Shrinkage reduces the mechanical interlocking and thus the fiber/matrix bond.

7.4 Toughness of CMCs

Many concepts have been proposed for augmenting the toughness of ceramic matrix materials [21]. Table 7.5, summarizes some of these concepts and gives the basic requirements for the models to be valid. Clearly, more than one toughness mechanism may be in operation at a given time. Matrix microcracking, fiber/matrix debonding leading to crack deflection and fiber fullout, and phase transformation toughening are all basically energy-dissipating processes that can result in an increase in toughness or work of fracture.

If the crack growth can be impeded by some means, then a higher stress would be required to make it move. Fibers (metallic or ceramic) can play the role of toughen-

Table 7.5. Ceramic matrix composite toughening mechanisms

Mechanism	Requirement
1. Compressive prestressing of the matrix	$\alpha_f > \alpha_m$ will result in an axial compressive prestressing of the matrix after fabrication
2. Crack impeding	Fracture toughness of the second phase (fibers or particles) is greater than that of matrix locally. Crack is either arrested or bows out (line tension effect).
3. Fiber (or whisker) pullout	Fibers or whiskers having high transverse fracture toughness will cause failure along fiber/matrix interface leading to fiber pullout on further straining.
4. Crack deflection	Weak fiber/matrix interfaces deflect the propagating crack away from the principal direction.
5. Phase transformation toughening	The crack tip stress field in the matrix can cause the second phase particles (fibers) at the crack tip to undergo a phase transformation causing expansion ($\Delta V > 0$). The volume expansion can *squeeze* the crack shut.

ing agents in ceramic matrices. Metallic fiber reinforced ceramics will clearly be restricted to lower temperatures than ceramic fiber reinforced ceramic matrices. Glass and glass-ceramic matrices containing carbon fibers [4–6, 12–14] have been shown to have fiber pullout as the dominant toughening mechanism. Basically, this mechanism, shown in Fig. 7.10, requires that the strength transferred to the fiber during the ceramic matrix fracture be less than the fiber ultimate strength, σ_{fu}, and that an interfacial shear stress be developed that is greater than the fiber/matrix interfacial strength, τ_i; that is, the interface must fail in shear. For a given fiber radius r_f, we have for the axial tensile stress in the fiber (see Chap. 10)

$$\sigma_f = 2\tau_i \left(\frac{l_c}{r_f}\right)$$

where l_c is the critical fiber length and $\sigma_f < \sigma_{fu}$. The tensile stress increases from a minimum at both fiber ends and attains a maximum along the central portion of the fiber (see Fig. 10.13). Fibers that bridge the fracture plane and whose ends terminate within $l_c/2$ from the fracture plane will suffer fiber pullout, while those with ends further away will fracture when $\sigma_f = \sigma_{fu}$. A crack deflection mechanism also requires a weak fiber/matrix bond so that as a matrix crack reaches the interface, it gets deflected along the interface rather than passing straight through the fiber. This is illustrated in Fig. 7.15 [22]. The original state involving frictional gripping of the fiber by the matrix is shown in Fig. 7.15a. On stressing the composite, a crack initiates in the matrix and starts propagating in the matrix normal to the interface. As it approaches the interface, the crack is momentarily halted by the fiber; see Fig. 7.15b. If the fiber/matrix interface is weak, then interfacial shear and lateral contraction of fiber and matrix will result in debonding and crack deflection away from its principal direction (normal to the interface); see Fig. 7.15c. A further increment of crack extension in the principal direction will occur after some delay. On continuing stressing of the composite, the fiber/matrix interface delamination continues and fiber failure will occur at some weak point along its length; see Fig. 7.15d. This is followed by broken fiber ends beings pulled out against the frictional resistance of the interface and finally causing a total separation.

DiCarlo [23] has discussed the requirements for strong and tough CMCs. Avoiding processing related flaws in the matrix and in the fiber would appear to be

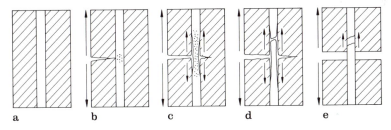

a   b   c   d   e

Fig. 7.15. a Original state with frictional gripping of the fiber by the matrix; **b** crack in the matrix is momentarily halted by the fiber; **c** interfacial shear and lateral contraction of fiber and matrix result in debonding and crack deflection along the interface; **d** further debonding, fiber failure at a weak point, and further extension; **e** broken fiber ends are pulled out against frictional resistance of interface, leading to total separation. (Reprinted from Ref. 22, used with permission.)

an elementary and straightforward recommendation. If processing results in large flaws in the matrix, the composite fracture strain will be low. In this respect, fiber bridging of cracks in a CMC will result in a reduced flaw size in the matrix. This in turn will help in achieving higher applied strains before crack propagation in the matrix than in an unreinforced, monolithic ceramic [24]. A weak interfacial bond, as pointed out above, leads to crack deflection at the fiber/matrix interface and/or fiber pullout. Use of a high volume fraction of continuous fibers, stiffer than the matrix, will give increased stiffness, which results in a higher stress level being needed to produce matrix microcracking and a higher composite ultimate tensile stress as well as a high creep resistance. A high volume fraction and a small fiber diameter also provide a sufficient number of fibers for crack bridging and postpone crack propagation to higher strain levels. A small-diameter fiber also translates into a small l_c, the critical length for effective load transfer from matrix to fiber (see Chap. 10). Although a weak interface is desirable from a toughness point of view, it provides a short circuit for environmental attack. Thus, an ability to maintain a high strength level and high inertness at high temperatures and in aggressive atmospheres is highly desirable.

7.5 Applications of CMCs

An important commercial application of CMCs is in cutting tools where they are replacing the metallic carbide cutting tools. Greenleaf Corp. (Saegertown, PA) makes a hot-pressed SiC_w/alumina for use as a cutting tool material. Figure 7.16 shows the microstructure of this composite. Potentially, CMCs can find applications in heat engines, components requiring resistance to aggressive environments, special electronic/electrical applications, energy conversion, and military systems [25]. Among the barriers that need to be overcome for large-scale applications of

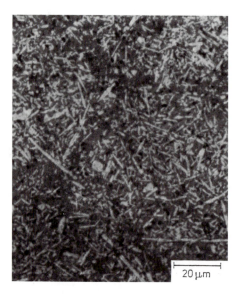

20 μm

Fig. 7.16. Microstructure of a SiC_w/Al_2O_3 composite cutting tool bit

CMCs are the high production costs, on accepted design philosophy, and the lack of models for strength and toughness [26]. Complicated shapes are difficult to make economically by hot pressing. Sintering or sintering followed by hot isostatic pressing (HIP) are the alternate routes for nonglassy matrices. Densifying ceramic bodies containing a reasonable volume fraction of fibers that inhibit sintering is indeed a formidable problem.

References

1   D.C. Phillips, in *Fabrication of Composites*, North-Holland, Amsterdam, 1983, p. 373.
2   J.A. Cornie, Y.-M. Chiang, D.R. Uhlmann, A. Mortensen, and J.M. Collins, *Am. Ceram. Soc. Bull.*, **65**, 293 (1986).
3   R.A.J. Sambell, D.C. Phillips, and D.H. Bowen, in *Carbon Fibres: Their Place in Modern Technology*, The Plastics Institute, London, 1974, p. 16/9.
4   K.M. Prewo and J.J. Brennan, *J. Mater. Sci.*, **15**, 463 (1980).
5   J.J. Brennan and K.M. Prewo, *J. Mater. Sci.*, **17**, 2371 (1982).
6   K.M. Prewo, J.J. Brennan, and G.K. Layden, *Am. Ceram. Soc. Bull.*, **65**, 305 (1986).
7   E. Fitzer and D. Hegen, *Angew. Chem.*, **91**, 316 (1979).
8   E. Fitzer and J. Schlichting, *Z. Werkstofftech.*, **11**, 330 (1980).
9   E. Fitzer and R. Gadow, *Am. Ceram. Soc. Bull.*, **65**, 326 (1986).
10 D.P. Stinton, A.J. Caputo, and R.A. Lowden, *Am. Ceram. Soc. Bull.*, **65**, 347 (1986).
11 A.G. Evans, *Mater. Sci. Eng.*, **71**, 3 (1985).
12 R.W. Davidge, *Mechanical Behavior of Ceramics*, Camb. U. Press, Cambridge, 1979, p.116.
13 D.C. Phillips, R.A.J. Sambell, and D.H. Bowen, *J. Mater. Sci.*, **7**, 1454 (1972).
14 K.M. Prewo, *J. Mater. Sci.*, **17**, 3549 (1982).
15 M. Herron and S.H. Risbud, *Am. Ceram. Soc. Bull.*, **65**, 342 (1986).
16 P.F. Becher and G.C. Wei, *Comm. Am. Ceram. Soc.*, **67**, 259 (1984).
17 G.C. Wei and P.F. Becher, *Am. Ceram. Soc. Bull.*, **64**, 298 (1984).
18 T.N. Tiegs and P.F. Becher, in *Tailoring multiphase and Composite Ceramics*, Plenum Press, New York, 1986, p. 639.
19 A.H. Chokshi and J.R. Porter, *J. Am. Ceram. Soc.*, **68**, c144 (1985).
20 R.A.J. Sambell, D.H. Bowen, and D.C. Phillips, *J. Mater. Sci.*, **7**, 773 (1972).
21 National Materials Advisory Board, *High Temperature Metal and Ceramic Matrix Composites for Oxidizing Atmosphere Applications*, NMAB-376, Washington, DC, 1981.
22 B. Harris, *Met. Sci.*, **14**, 351 (1980).
23 J.A. DiCarlo, *J. Met.* **37**, 44 (June 1985).
24 J. Aveston, G.A. Cooper, and A. Kelly, in *The Properties of Fibre Composites*, IPC Science & Technology Press, Guildford, England, 1971, p. 15.
25 L.J. Schioler and J.J. Stiglich, *Am. Ceram. Soc. Bull.*, **65**, 289 (1986).

Suggested Reading

American Ceramic Society Bulletin, **65** (Feb. 1986): a special issue on ceramic matrix composites.
R.F. Davis, H. Palmour III, and R.L. Porter (eds.), *Emergent Process Methods for High-Technology Ceramics*, Plenum Press New York, 1984.
D.C. Phillips, Fiber Reinforced Ceramics, in *Fabrication of Composites*, vol. 4 of *Handbook of Composites*, North-Holland, Amsterdam, 1983, p. 373.
R.E. Tressler and G.L. Messing (eds.), *Tailoring Multiphase and Composite Ceramics*, Materials Science Research Series, Plenum Press, New York, 1986.

8. Carbon Fiber Composites

Carbon fiber composites started out in the 1950s and attained the status of a mature structural material in the 1980s. Not unexpectedly, the aerospace industry has been the biggest user of carbon fiber reinforced polymer matrix composites, followed by the sporting goods industry. The availability of a large variety of carbon fibers (Chap. 2) and an equally large variety of polymer matrix materials (Chap. 3) made it easier for carbon fiber reinforced polymer matrix composites to assume the important position that they have. This is the reason we devote a separate chapter to this class of composites. Epoxy is the most commonly used polymer matrix with carbon fibers. Polyester, polysulfone, polyimide, and thermoplastic resins are also used. Carbon fibers are the major load-bearing components in most such composites. There is, however, a class of carbon fiber composites wherein the excellent electrical conduction characteristics of carbon fibers are exploited; for example in situations where static electric charge accumulation occurs, parts made of thermoplastics containing short fibers are frequently used. As we did for other composite systems, we describe the fabrication, properties, interfaces, and applications of carbon fiber reinforced polymer matrix composites. A special emphasis is given to carbon/carbon composites, an important subclass.

8.1 Fabrication of Carbon Fiber Composites

Most fabrication methods described in Chap. 5 for polymer matrix composites (PMCs), such as pultrusion, vacuum molding, filament winding, and laminated composites starting from prepregs, are also used for carbon fiber reinforced PMCs. An example of a product obtained by pultrusion is shown in Fig. 8.1. The hollow trapezoidal shaped product shown is a helicopter windshield post made of carbon fiber mat and tows in a high-temperature vinyl ester resin matrix. Injection or compression molding of chopped carbon fibers in a thermoplastic or thermoset resin is an economical technique, especially where mechanical property requirements are not very critical. In the aerospace industry, carbon fiber composites are generally made from prepreg sheets or tapes that allow a high fiber volume fraction to be attained. Prepregs are rolls of sheets of unidirectionally oriented fibers preimpregnated with a partially cured resin matrix. A removable backing sheet is used to prevent sticking of prepregs. Prepregs are stacked in an appropriate sequence (see Chap. 11) and consolidated into a composite component in an autoclave (see Chap. 5).

Fig. 8.1. A helicopter windshield post made of carbon fibers/vinyl ester resin by pultrusion. The post is 1.5 m long. (Courtesy of Morrison Molded Fiber Glass Co.)

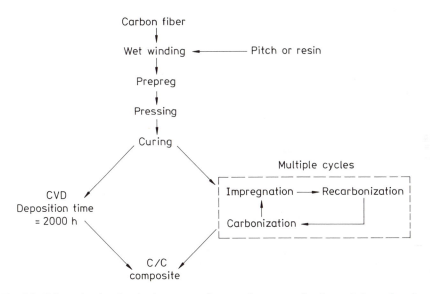

Fig. 8.2. Schematic of carbon/carbon composite manufacture starting from pitch or phenolic resin. (From Ref. 1, used with permission.)

Carbon fiber reinforced carbon matrix composites, or the so-called carbon/carbon composites (C/C), do require different processing techniques. These are actually closer to ceramic matrix composites than polymer matrix composites. There are three main routes to obtain carbon/carbon composites:

1. A woven carbon fiber preform is impregnated under heat and pressure with pitch from coal tar or petroleum sources. This is followed by pyrolysis. The cycle may be repeated to obtain the desired amount of densification.
2. Carbon fiber/polymer composites are pyrolyzed to decompose the resin, generally phenolics because they give high char strength, followed by reimpregnation and repyrolysis to get a carbonaceous matrix bonded to carbon fibers.
3. Chemical vapor deposition from a gaseous phase (methane + nitrogen + hydrogen) onto and between the carbon fibers in a preform.

A schematic of the phenolic or pitch route for obtaining carbon/carbon composites is shown in Fig. 8.2. The starting material may be a carbon/fiber phenolic

prepreg or a three-dimensional woven structure, say a cone, which can be impregnated under pressure with resin or pitch. In the case of prepregs, they are cut and laid
up in a form appropriate for the final desired shape. Application of heat and
pressure results in curing of the matrix and fixing of the carbon fibers in the desired
shape. This is followed by a densification cycle involving carbonization in an inert
atmosphere (an optional high-temperature graphitization treatment is also sometimes used) and reimpregnation by resin or pitch to obtain the desired density. The
densification cycle, indicated by the dashed lines in Fig. 8.2, may be repeated several
times to reduce the porosity to an acceptable level. This densification may also be
carried out by CVD of carbon from a hydrocarbon gas (e.g., methane, acetylene, or
benzene). A characteristic feature of pyrolysis of the polymer or pitch infiltrants is
the rather heavy weight loss, anywhere between 10 and 60% which is inevitably
accompanied by shrinkage porosity. Fritz et al. [1] and McAllister and Lachman [2]
give details of the carbon/carbon composite manufacturing processes.

8.2 Properties

Since a variety of carbon fibers is available, one can produce carbon fiber reinforced composites showing a range of properties. Two major types of carbon fiber are
the high-strength and high-modulus types (Sect. 2.4). Accordingly, we summarize
some of the mechanical characteristics of unidirectionally aligned, 62% volume
fraction, carbon fibers from different sources in an epoxy matrix in Table 8.1 [3].

The carbon fibers find a major outlet in the aerospace industry. Because aerospace structures are exposed to a range of environments and temperatures, for

Table 8.1. Typical mechanical properties of some carbon fiber/epoxy composites[a]

Property	AS	HMS	Celion 6000	GY 70
Tensile strength (MPa)	1850	1150	1650	780
Tensile modulus (GPa)	145	210	150	290–325
Tensile strain to fracture (%)	1.2	0.5	1.1	0.2
Compressive strength (MPa)	1800	380	1470	620–700
Compressive modulus (GPa)	140	110	140	310
Compressive strain to fracture (%)	—	0.4	1.7	—
Flexural strength (4 points) (MPa)	1800	950	1750	790
Flexural modulus (GPa)	120	170	135	255
Interlaminar shear strength (MPa)	125	55	125	60

[a]Values given are indicative only and for a unidirectional composite (62% V_f) in the longitudinal
direction.
Source: Adapted with permission from Ref. 3.

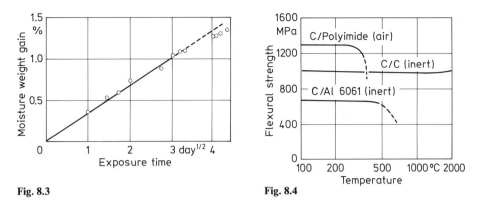

Fig. 8.3   **Fig. 8.4**

Fig. 8.3. Fickian moisture absorption in a carbon/epoxy laminate. (Adapted from Ref. 4, used with permission.)

Fig. 8.4. Comparison of flexural strength of carbon/polyimide (in air) and carbon/carbon and carbon/aluminum (insert atmosphere) as a function of temperature. (From Ref. 6, used with permission.)

example, oils, fuels, moisture, acids, and hot gases, the excellent corrosion resistance characteristics of carbon/epoxy composites are of great value under such conditions. Commonly encountered damage to polymers by ultraviolet rays is minimized by properly painting the exterior of the composite. Moisture is a major damaging agent. Epoxy matrices can absorb water to as much as 1% of the composite weight; however, unlike glass fiber which is attacked by moisture, the carbon fiber itself is unaffected by moisture. Thus, moisture absorbed in carbon fiber PMCs opens up the polymer structure and reduces its glass transition temperature; that is, the moisture acts as a plasticizer of the polymeric matrix. Moisture absorption in polymers occurs according to Fick's law; that is, the weight gain owing to moisture intake varies as the square root of the exposure time. This Fickian moisture absorption in a 16-ply carbon epoxy laminate is shown in Fig. 8.3 [4]. The moisture absorption problem in PMCs is generally related to that of degradation by temperature effects. Collings and Stone [5] present a theoretical analysis of strains developed in longitudinal and transverse plies of a carbon/epoxy laminate owing to hygrothermal effects. An interesting finding of theirs is that tensile thermal strains that develop in the matrix after curing are reduced by compressive strains generated in the matrix by swelling resulting from water absorption. It is worth pointing out that moisture absorption causes compressive stresses in the resin and tensile stresses in the fibers. Also, a temperature increase ΔT generates strains of the same sign as those caused by an increase in the moisture content [5]. Thus, it is understandable that moisture absorption should reduce the residual strains after curing. In view of the high-temperature resistance requirements, it is instructive to look at the high-temperature strength of carbon fiber composites. Figure 8.4 shows flexural strengths of carbon/polyimide in air, and those of carbon/carbon and carbon/aluminum in an inert atmosphere, as a function of temperature [6]. While carbon/polyimide shows a superior high-temperature strength than most carbon fiber PMCs, we note that carbon/aluminum shows good strength up to about 500°C and that carbon/carbon can withstand temperatures as high as 2000°C. It is worth point-

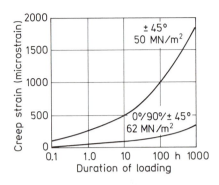

Fig. 8.5. Creep strains at ambient temperature for a ± 45 and for a $0°/90°/\pm 45°$ carbon/epoxy laminate. (From Ref. 8, used with permission.)

ing out that the excellent high-temperature characteristics of carbon/aluminum and carbon/carbon are obtainable in inert atmospheres only. In air, carbon fibers are readily oxidized and protective coatings are needed.

Carbon fiber/epoxy composites exhibit superior properties in creep compared to Kevlar 49/epoxy. Kevlar 49 creeps significantly even at quite low stresses [7]. The ply stacking sequence can affect the composite properties. Figure 8.5 shows tensile creep strain at ambient temperature as a function of time for two different stacking sequences [8]. Note that a laminate with carbon fibers at $\pm 45°$ shows more creep strain than one containing plies at $0°/90°/\pm 45°$. The reason for this is that in the $\pm 45°$ sequence, the epoxy matrix creeps by (a) tension in the loading direction, (b) shear in the $\pm 45°$ directions, and (c) rotation of the plies in a scissorlike action. As we shall see in Chap. 11, the addition of $0°$ and $90°$ plies eliminates the scissorlike rotation characteristic of $45°$ plies and reduces the matrix shear deformation. Thus, for creep resistance, the $0°/90°/\pm 45°$ sequence is to be preferred over the $\pm 45°$ sequence.

Carbon fiber PMCs generally show excellent fatigue strength. Depending on the ply stacking arrangement, their fatigue strength (tension–tension) may vary from 60 to 80% of the ultimate tensile strength for lives over 10^7 cycles [9]. The higher fatigue strength levels pertain to composites having more than 50% of the fibers in the longitudinal direction ($0°$), which leads to high longitudinal stiffness and low strains. Pipes and Pagano [10] showed that certain stacking sequences can result in tensile stresses at the free edges, which can lead to early local delamination effects in fatigue and consequently to lower fatigue lives. This is discussed in Chaps. 11 and 12.

One of the areas where extensive efforts have been made is that of toughening the carbon fiber PMCs. This has involved modifying epoxies and using polymeric matrix materials other than epoxies. Among the latter are modified bismaleimides and some new thermoplastic materials (see Chap. 3 also). The latter category includes poly (phenylene sulfide) (PPS), polysulfones (PS), and polyetheretherke-tone (PEEK) among others. PEEK, a semicrystalline polyether, combines excellent toughness with chemical inertness. Typically, PEEK-based carbon fiber composites are equivalent to high-performance epoxy-based carbon fiber composites but with the big advantage that the former have an order of magnitude higher toughness than the latter. PPS is a semicrystalline aromatic sulfide that has excellent properties. Phillips Petroleum has developed a special process of making continuous fiber prepregs with this resin for use in making composite laminates.

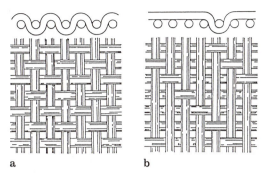

Fig. 8.6. Two-dimensional fabric: **a** plain weave; **b** five-harness satin. (Courtesy of Fiber Materials Inc.)

a   b

Table 8.2. Mechanical properties of carbon/carbon composites at room temperature

Weave	Flexural strength (MPa)	Young's modulus (GPa)	Interlaminar shear strength (MPa)
1-D, 55% V_f	1200–1400	150–200	20–40
2-D, 8H/S weave,			
35% V_f	300	60	20–40
3-D, Felt, 35% V_f	170	15–20	20–30

Source: Adapted with permission from Ref. 1.

Properties of carbon/carbon composites depend on the type of carbon fiber used (high-modulus or high-strength type), fiber volume fraction, the fiber distribution. One-, two-, and three-dimensionally woven carbon fibers may be used. Figure 8.6 shows a two-dimensional (2-D) plain weave and a five-harness satin weave [2]. Modifications of the basic 3-D orthogonal weave involving 5, 7, or 11 fiber directions are possible, which give a highly isotropic final composite [2]. Klein [11] has described the variety of weaves available for reinforcement in two or more dimensions. A plain weave, for example, has one warp yarn running over and under one fill yarn and is the simplest weave. Satin type weaves are more flexible; that is, they can conform to complicated shapes easily. A five-harness satin weave shown in Fig. 8.6b has one warp yarn running over four fill yarns and under one fill yarn. It should be mentioned that the carbonaceous matrix is not an unimportant factor. The precursor, processing, and the high temperature involved all influence the matrix significantly. Table 8.2 summarizes the room temperature properties of some carbon/carbon composites. The values refer to a high-modulus carbon fiber, a final heat treatment at 1000°C, four to six densification cycles, and a fiber volume fraction of 55. The properties of a 2-D weave depend on the type of weave of carbon cloth. The properties depend strongly on the weave pattern and the amount of fibers in the x, y, and z directions. Figure 8.7 shows schematically the stress–strain curves, at room temperature, of 1-D, 2-D, and 3-D carbon/carbon composites [1]. Note the change in fracture mode from semibrittle (1-D) to nonbrittle (3-D). The latter is due to the existence of the continuous crack pattern in the composite. Since carbon/carbon composites are meant for high-temperature applications, it is of

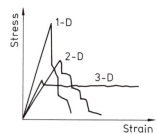

Fig. 8.7. Stress–strain curves (schematic) of 1-D, 2-D, and 3-D carbon/carbon composites. (From Ref. 1, used with permission.)

interest to examine their thermal expansion behavior. As expected, fibers control the thermal expansion behavior parallel to the fibers, while perpendicular to the fiber axis the carbonaceous matrix controls the expansion behavior. The amount of porosity in the matrix will also influence the thermal expansion behavior.

8.3 Interface

The interface region between a carbon fiber and the polymer matrix is quite complex. It is therefore not surprising that a unified view of the interface in such composites does not exist. As pointed out in Chap. 2, the carbon fiber structure is not homogeneous through its cross section. The orientation of the basal planes depends on the precursor fiber and processing conditions. In particular, the so-called onion-skin structure is frequently observed in PAN-based fibers, wherein basal planes in a thin surface layer are aligned parallel to the surface. Figure 8.8 shows the structure of a carbon fiber/epoxy composite [12]. The onion-skin zone (C in Fig. 8.8) has a very graphitic structure and is quite weak in shear. Thus, failure is likely to occur in this thin zone. Additionally, the skin can be hard to bind with a polymeric matrix because of the high degree of preferred orientation of the basal planes, thus, facilitating interfacial failure (zone D). The matrix properties in zone

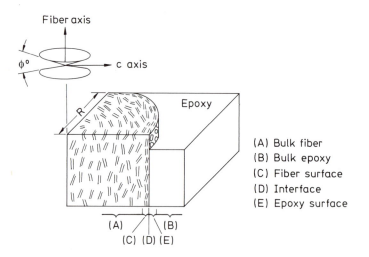

(A) Bulk fiber
(B) Bulk epoxy
(C) Fiber surface
(D) Interface
(E) Epoxy surface

Fig. 8.8. Structure of carbon fiber, interface, and epoxy matrix. (From Ref. 12, used with permission.)

E (close to the interface) may be different from those of the bulk epoxy (zone B). Carbon fibers meant for polymer reinforcement invariably receive some form of surface treatment from the manufacturer to improve their compatibility with the polymer matrix as well as their handleability. Organic "sizes" are commonly applied by passing the fibers through a sizing bath. Common sizes include polyvinyl alcohol, epoxy, polyimide, and titanate coupling agents.

Carbon fibers, especially high-modulus carbon fibers that have undergone a high-temperature graphitization, are quite smooth. They have a rather low specific area, varying from 0.1 to 2 $m^2 g^{-1}$. The small amount of roughness is present mostly as longitudinal striations (see Fig. 2.19). Carbon fibers are also generally chemically inert; that is, interfacial interactions in carbon-fiber-based composites would be rather weak. Generally, a short beam bending test is conducted to measure what is called interlaminar shear strength (ILSS). Admittedly, such a test is not entirely satisfactory, but for lack of any better, quicker, or more convenient test, the ILSS test value is taken as a measure of bond strength.

Ehrburger and Donnet [13] point out that there are two principal ways of improving interfacial bonding in carbon fiber composites: (a) increase the fiber surface roughness, and thus the interfacial area, and (b) increase the surface reactivity. Many surface treatments have been developed to obtain improved interfacial bonding between carbon fibers and the polymer matrix [14, 15]. A brief description of these follows.

8.3.1 Chemical Vapor Deposition (CVD)

SiC and pyrocarbon have been deposited on carbon fibers by CVD. Growing whiskers on the carbon fiber surface, called whiskerization in the literature, can result in a two- to threefold increase of ILSS. SiC whiskers (~ 2 μm diameter) are grown on the fiber surface. The improvement in ILSS is mainly due to the increase in surface area. Whiskerizing involves the growth of single-crystal SiC whiskers perpendicular to the fiber surface, which results in an efficient mechanical keying effect with the polymer matrix. The CVD method is expensive and handling whiskerized fibers is difficult. Although this treatment results in an increase in ILSS, there also occurs a weakening of the fibers.

8.3.2 Oxidative Etching

Treating carbon fibers with several surface-oxidation agents leads to significant increases in the ILSS of composites. This is because the oxidation treatment increases the fiber surface area and the number of surface groups [13]. Yet another reason for increased ILSS may be the removal of surface defects, such as pores, weakly bonded carbon debris, and impurities [14]. Oxidation treatments can be carried out by a gaseous or a liquid phase. Gas phase oxidation can be done with air or oxygen diluted with an inert gas [16]. Gas flow rates and the temperature are the important parameters in this process. Oxidation results in an increase in the fiber surface roughness by pitting and increased longitudinal striations [14]. Too high an oxidation rate will result in nonuniform etching of the carbon fibers and a loss of fiber tensile strength. Since oxidation results in a weight loss, we can conveniently

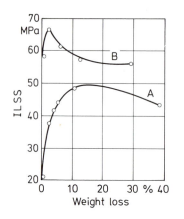

Fig. 8.9. Interlaminar shear strength (ILSS) as a function of weight loss on oxidation. A high weight loss corresponds to a high degree of oxidation. Note that the maximum ILSS corresponds to less than a 10% weight loss in both cases. A: High-modulus type; B: High-strength type. (From Ref. 16, used with permission.)

take the amount of weight loss as an indication of the degree of oxidation. Figure 8.9 shows ILSS versus weight loss for high-strength and high-modulus carbon fiber composites [17]. The maximum ILSS corresponds to less than a 10% weight loss in both cases. Overoxidation results in a loss of fiber strength and lower ILSS.

8.3.3 Liquid Phase Oxidation

Liquid phase oxidation involves treatments in nitric acid, sodium hypochlorite, potassium permanganate, and anodic etching [14]. Liquid phase oxidation by nitric acid and sodium hypochlorite results in an increase in interfacial area and formation of oxygenated surface groups due to fiber etching. Wetting of the carbon fibers by the polymer is enhanced by these changes.

Graphitic oxides are lamellar compounds having large amounts of hydroxylic and carboxylic groups. The formation of a graphitic oxide layer increases the number of acidic groups on the carbon fibers.

Anodic etching or electrochemical oxidation using dilute nitric acid or dilute sodium hydroxide solutions results in no significant decrease in tensile strength of the carbon fibers, according to Ehrburger and Donnet [13]. The fiber weight loss is less than 2% and no great change in surface area or fiber roughness occurs, the major change being an increase in the acidic surface groups. Oxidative treatments produce functional groups ($-CO_2H$, $-C-OH$, and $-C=O$) on carbon fiber surfaces. They form at the edges of basal planes and at defects. These functional groups form chemical bonds with unsaturated resins.

Drzal et al. [18, 19] studied the role of carbon fiber surface treatments. Their conclusion was that various carbon fiber surface treatments promoted adhesion to epoxy materials through a two-pronged mechanism: (a) the surface treatment removes a weak outer layer that is present initially on the fiber and (b) chemical groups are added to the surface which increase interaction with the matrix. Contrary to other researchers, these authors did not find an increase in the surface area to be important in promoting fiber/matrix adhesion. When a fiber finish is applied, according to Drzal and coworkers, the effect is to produce a brittle but high-modulus interphase layer between the fiber and the matrix. Their model is shown in Fig. 8.10. Initially, the fiber has a surface finish containing 0 parts per hundred of

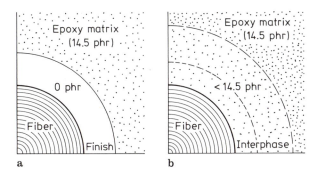

Fig. 8.10. Model of interphase development in a carbon fiber/ epoxy matrix: **a** initial, **b** final. (From Ref. 19, used with permission.)

amine (0 phr). The epoxy matrix has about 14.5 parts per hundred of amine (14.5 phr). Upon curing, the amine from the bulk matrix diffuses into the finish layer and an interphase region is formed whose properties show a gradient from the fiber surface out. According to Drzal and coworkers, as amine content is reduced, the modulus of epoxy goes up accompanied by lower fracture strength and strain. That is, the interphase being created between the fiber and the matrix has high modulus but low toughness. This is supposed to promote matrix fracture as opposed to interfacial fracture.

8.4 Applications

The aerospace industry is the major user of lightweight structures made of carbon fiber reinforced polymer matrix composites. Carbon fiber PMCs started entering the aircraft industry as early as the 1970s. In commercial aircraft, such as the Boeing 757 and 767, carbon fiber composites are used in a number of locations. The weight savings achieved in the Boeing 767 through the use of composites amounted to about 1000 kg over conventional metallic structures. An aircraft louvered door made of carbon fiber (40%) in a nylon 6/12 matrix that is used on the engine nacelle of the Boeing 757 aircraft is shown in Fig. 8.11. The helicopter industry in particular has been an enthusiastic user of carbon fiber PMCs. Figure 8.1 shows a helicopter windshield post made of carbon fiber/vinyl ester. It is claimed that composite helicopter rotor blades have lower direct operating costs than aluminum blades [20], and this is without taking into account the intangibles such as longer fatigue life, reduced maintenance, and lower manufacturing costs. Boeing Vertol, Bell Aircraft,

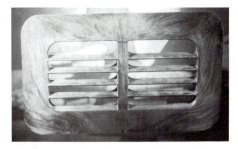

Fig. 8.11. A louvered door made of carbon fiber (40% V_f)/nylon thermoplastic used on the engine nacelle of the Boeing 757 aircraft. (Courtesy of LNP Corporation.)

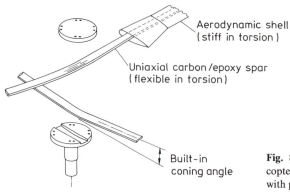

Aerodynamic shell
(stiff in torsion)

Uniaxial carbon/epoxy spar
(flexible in torsion)

Built-in
coning angle

Fig. 8.12. Schematic carbon/epoxy heli-copter rotor blades. (From Ref. 20, used with permission.)

Fig. 8.13   **Fig. 8.14**

Fig. 8.13. Intelsat 5 has a primary structure as well as several antennas made of carbon fiber PMCs. (Courtesy of Fiberite Co.)

Fig. 8.14. Examples of sporting goods made of carbon fiber composites where lightness, good mechanical characteristics, and sleek lines make the items very attractive. (Courtesy of Fiberite Co.)

and Sikorsky, among others, use composite rotor blades in their helicopters. Figure 8.12 shows a schematic of a carbon/epoxy helicopter rotor blade. Cargo bay doors, maneuvering engine pods, arm booms, and booster rocket casings in the U.S. Space Shuttle Orbiter are made of carbon fiber/epoxy composites. Figure 8.13 shows the primary tower structure and several antennas made of carbon fiber/epoxy for use in Intelsat 5. The main attractions for their use are lightness and dimensional stability.

In the road transport industry carbon fiber/epoxy composites have been tried in leaf springs, driveshafts, and various chassis parts. Blades in turbines, compressors, and windmills, as well as in ultracentrifuges and flywheels, have been made of carbon fiber PMCs.

Medical applications of carbon fiber composites have involved their use as prosthetic devices, as well as in the manufacture of medical equipment, for example,

X-ray film holders and tables for X-ray equipment. The ability to tailor the required stiffness and strength properties of internal bone plates made of carbon fiber composites gives them a definite advantage over metallic parts.

After the aerospace industry, it is in the leisure and sporting goods industry where carbon fiber PMCs have found the greatest number of applications. Golf clubs, archery bows, fishing rods, tennis rackets, cricket bats, and skis are common-place items where carbon fiber PMCs are used. Figure 8.14 shows a tennis racket, a ski, and a fishing rod made of carbon fiber PMCs. Note the sleek lines that have become a hallmark of composite construction. In addition to the fine mechanical characteristics, carbon fiber reinforced thermoplastic composites have excellent electrical properties. This is exploited in situations where a static charge builds up easily, for example, in high-speed computer parts and in musical instruments where rubbing, sliding, or separation of an insulating material results in electrostatic voltages. In parts made of insulating polymeric resins, this charge stays localized until the polymer comes in contact with a body at a different potential and the electrostatic voltage discharges via an arc or spark. Voltages as high as 30–40 kV can build up. This electrostatic charge can be painful to a human being or even fatal in extreme circumstances. Some very sensitive microelectronic parts can be damaged by an electrostatic discharge of a mere 20 V. Thus, it is not surprising that carbon fibers dispersed in thermoplastic resins are finding applications where dissipation of a static charge is important. Conductive fillers of one kind or another were used earlier but carbon fibers also serve to reinforce mechanically in situations requiring high strength and wear resistance. Figures 8.15a and b show a microphone and head shell, the unit that holds the stylus at the end of the turntable arm. Carbon fibers dissipate the static directly, thus, eliminating the need of copper conductors. This is coupled with a high stiffness to weight ratio of these composites which allows a weight reduction of the part as well. Shielding against electromagnetic interference (EMI) is another area where highly conductive composites based on carbon fibers are finding applications. EMI is nothing but electronic pollution or noise caused by rapidly changing voltages. Examples include avionic housings, computer enclosures, and any other electronic device that needs protection against

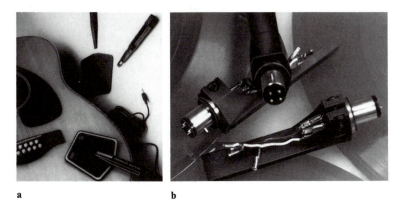

a   b

Fig. 8.15. Use of carbon/thermoplastic composites in situations involving static charge: **a** a microphone, **b** a head shell unit of a turntable arm. (Courtesy of LNP Corporation.)

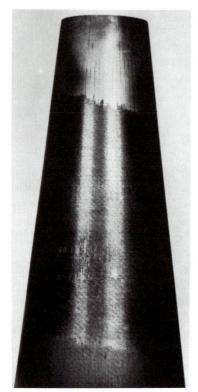

Fig. 8.16. A fully processed three-dimensional carbon/carboon frustum. (Courtesy of Fiber Materials Inc.)

stray EMI. While there are applications involving the use of the conductive properties of carbon fibers, there are also problems associated with this characteristic of carbon fibers. Carbon fibers are extremely fine and light and if they become accidently airborne, for any reason, and settle in on an electrical equipment, short circuiting can occur. NASA did extensive studies on this problem and concluded that although the risk was there, it did not warrant prohibiting the use of carbon fibers in structures [21].

Major applications of carbon/carbon composites involve uses at high temperatures, for example, as heat shields for reentry vechicles, aircraft brakes, hot-pressing dies, and high-temperature parts such as nozzles. Figure 8.16 shows a fully processed three-dimensional carbon/carbon frustum. Hot-pressing dies made of carbon/carbon composites are commercially available while brake disks are used in military aircraft, the Concorde supersonic aircraft, and racing cars. Heat shields and nozzles are generally made of multidirectionally reinforced carbon/carbon composites. Figure 8.4 compares the flexural strength of various carbon fiber composites. Note that carbon/carbon composites can withstand high temperatures in an inert atmosphere. Lack of oxidation resistance is a major problem and a great deal of effort is being put into the development of oxidation-resistant coatings for carbon fibers, with SiC coatings being a primary candidate.

References

1   W. Fritz, W. Huttner, and G. Hartwig, in *Nonmetallic Materials and Composites at Low Tempera-tures*, Plenum Press, New York, 1979, p. 245.
2   L.E. McAllister and W.L. Lachman, in *Fabrication of Composites*, North-Holland, Amsterdam, 1983, p. 109.
3   D.M. Riggs, R.J. Shuford, and R.W. Lewis, in *Handbook of Composites*, Van Nostrand Reinhold, New York, 1982, p. 196.
4   C.D. Shirrel and F.A. Sandow, in *Fibrous Composites in Structural Design*, Plenum Press, New York, 1980, p. 795.
5   T.A. Collings and D.E.W. Stone, *Composites*, **16**, 307 (1985).
6   E. Fitzer and M. Heym, *Chem. Ind.*, 663 (Aug. 21, 1976).
7   R.H. Eriksen, *Composites*, **7**, 189 (1976).
8   J.B. Sturgeon, in *Creep of Engineering Materials*, a Journal of Strain Analysis Monograph, 1978, p. 175.
9   A. Baker, *Met. Forum*, **6**, 81 (1983).
10 R.B. Pipes and N.J. Pagano, *J. Composite Mater.* **1**, 538 (1970).
11 A.J. Klein, *Adv. Mater. Proc.* **2**, 40 (Mar. 1986).
12 R.J. Diefendorf, in *Tough Composite Materials*, Noyes Publishing, Park Ridge, NJ, 1985, p. 191.
13 P. Ehrburger and J.B. Donnet, *Philos. Trans. R. Soc. London*, **A294**, 495 (1980).
14 J.-B. Donnet and R.C. Bansal, *Carbon Fibers*, Marcel Dekker, New York, 1984, p. 109.
15 D. Mackee and V. Mimeault, in *Chemistry and Physics of Carbon*, vol. 8, Marcel Dekker, New York, 1973, p. 151.
16 D. Clark, N.J. Wadsworth, and W. Watt, in *Carbon Fibres, Their Place in Modern Technology*, The Plastics Institute, London, 1974, p. 44.
17 F. Molleyre and M. Bastick, *High Temp. High Pressure*, **9**, 237 (1977).
18 L.T. Drzal, M.J. Rich, and P.F. Lloyd, *J. Adhesion*, **16**, (1983).
19 L.T. Drzal, M.J. Rich, M.F. Koenig, and P.F. Lloyd, *J. Adhesion*, **16**, 133 (1983).
20 N.J. Mayer, in *Engineering Applications of Composites*, Academic Press, New York, 1974, p. 24.
21 *Risk to the Public from Carbon Fibers Released in Civil Aircraft Accidents*, NASA SP-448, NASA, Washington DC, 1980.

Suggested Reading

A.A. Baker, *Met. Forum*, **6**, 81 (1983).
J. Delmonte, *Technology of Carbon and Graphite Fiber Composites*, Van Nostrand Reinhold, New York, 1981.
J.-B. Donnet and R.C. Bansal, *Carbon Fibers*, Marcel Dekker, New York, 1984.
P. Ehrburger and J.-B. Donnet in *Handbook of Fiber Science & Technology*, vol. III, *High Technology Fibers*, Part A, Marcel Dekker, New York, 1985, p. 169.
E. Fitzer, *Carbon Fibres and Their Composites*, Springer-Verlag, Berlin, 1985.

9. Multifilamentary Superconducting Composites

9.1 Introduction

Multifilamentary composite superconductors started becoming available in the 1970s. These are niobium based (Nb-Ti and Nb_3Sn) superconductors. The record high temperature at which a material became superconductor was 23 K and was set in 1974. In 1987, there started appearing reports of superconductivity at temperatures up to 90 or 100 K in samples containing lanthanum, copper, oxygen, and barium or another IIa metal. These new ceramic superconductors have layered perovskite body-centered tetragonal structure and, not surprisingly, are very brittle. There remains an extremely large gap to be bridged between producing a small sample for testing in the laboratory and making a viable commerical product. The Nb-Ti system took 15–20 years between the discovery and the commercial availability. The new high-temperature oxide superconductors hold a great promise and it is quite likely eventually they will also be made into some kind of composite superconductors. Thus, it is quite instructive to review the composite materials aspects of niobium based superconductors. But first a short introduction to the subject of superconductivity is in order.

Certain metals and alloys lose all resistance to the flow of electricity when cooled to within a few degrees of absolute zero. The phenomenon is called superconductivity and the materials exhibiting this phenomenon are called superconductors. These superconductors can carry a high-density current without any electrical resistance; thus, they can generate the very high magnetic fields that are common in high-energy physics and fusion energy programs. Other fields of application include magnetic levitation vehicles, magnetohydrodynamic generators, rotating machines, and magnets in general.

Superconductivity is a very complex phenomenon. Bardeen, Cooper, and Schrieffer won a Nobel prize for explaining this phenomenon. An extremely simple account of their theory is as follows. Electrons have negative charge. Therefore, under normal circumstances, one would expect any two electrons to repel each other. However, according to the Bardeen–Cooper–Schrieffer (BCS) theory of superconductivity, at extremely low temperatures there arises an attractive force between two electrons and they become bound together. These electron pairs, called Cooper pairs, are able to overcome the inherent repulsion that one would expect between any two particles carrying the same charge. Electrical resistance of a normal material results from the scattering of electron waves by lattice virbations. In the superconducting state, near absolute zero, the Cooper electron pairs move in

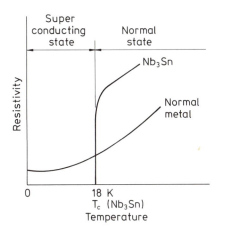

Fig. 9.1. Variation of electrical resistivity with temperature for a normal metal and a superconducting material, Nb_3Sn

phase. Thus, a change in the momentum of one pair needs a compensating change in all the other pairs. Under such conditions, the energy needed to scatter an electron pair and therefore to cause electrical resistance is much larger than the lattice vibrational energy available at such low temperatures. This simple account of the theory of superconductivity will suffice since our main focus in this chapter is on superconducting materials.

Kammerlingh Onnes discovered the phenomenon of superconductivity in 1911. Since then, some 27 elements and hundreds of solid solutions or compounds have been discovered that show this phenomenon of total disappearance of electrical resistance below a critical temperature T_c. Figure 9.1 shows the variation of electrical resistivity with temperature of a normal metal and that of a superconducting material, Nb_3Sn. T_c is a characteristic constant of each material. Kunzler and coworkers [1] in 1961 discovered the high critical field capability of Nb_3Sn and thus opened up the field of practical, high-field superconducting magnets. It turns out that most of the superconductors came into the realm of economic viability when techniques were developed to put the superconducting species in the form of ultrathin filaments in a copper matrix.

9.2 Types of Superconductor

The technologically most important property of superconductors is their capacity to carry an electric current without normal I^2R losses up to a critical current density, J_c. This critical current is a function of the applied field and temperature. The commercially available superconductors in the 1980s could demonstrate critical current densities of $J_c > 10^4$ A mm^{-2} at 4.2 K and an applied field of $H = 5$ T.

There are three parameters that limit the properties of a superconductor, namely, the critical temperature (T_c), the critical electrical current density (J_c), and the critical magnetic field (H_c); see Fig. 9.2. As long as the material does not cross into the shaded area indicated in Fig. 9.2, it will behave as a superconductor.

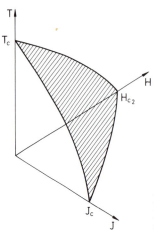

Fig. 9.2. The state of superconductivity is described by three critical parameters: magnetic field (H), temperature (T), and current density (J)

There are two types of superconductor:

Type I: These are characterized by low T_c values and they lose their superconductivity abruptly at H_c.

Type II: These behave as diamagnetic materials up to a field H_{c1}. Above this field, the magnetic field penetrates gradually into the material and concomitantly the superconductivity is gradually lost, until at the critical magnetic field H_{c2} the material reverts to the normal state. All major applications of superconductivity involve the use of these type II superconductors.

9.3 High-Field Composite Superconductors

The purpose of the superconductor being in the form of a filamentary composite is explained first, followed by a description of some important superconducting composite systems and their applications.

9.3.1 The Problem of Flux Pinning

Magnets for plasma confinement in a fusion reactor and large solenoids and coils for rotating machinery need rather long-length superconductors of uniform properties. These superconductors are in the form of a myriad of ultrathin filaments incorporated in a copper matrix to form a filamentary fibrous composite. Superconducting filaments have a micrometer-size diameter that helps to reduce the risk of flux jump in any given filament.

Any perturbance of a superconductor, say, by motion or a change in the applied field, leads to a rearrangement of magnetic flux lines in the superconductor. The phenomenon is called flux motion and is an energy-dissipative, that is, heat-producing, process. This temperature increase results in a reduced critical current

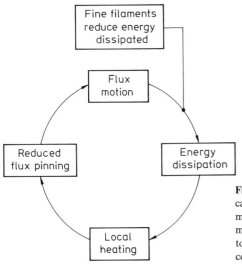

Fine filaments reduce energy dissipated

Flux motion

Reduced flux pinning

Energy dissipation

Local heating

Fig. 9.3. Flux movement in a superconductor is caused by any change in current of field by any mechanical disturbance. The end result of flux movement is that the superconductor is quenched to normal state. Filamentary composite superconductors help pin the flux.

and more flux motion results. The net result is that the superconductor is heated above T_c and reverts to the normal state. A practical solution to this problem is to make the superconductor in the form of ultrathin filaments so that the amount of energy (heat) dissipated by flux motion is too small to cause this runaway behavior; see Fig. 9.3. The high-purity copper metal provides a high-conductivity alternate path for the current. In the case of a *quench*, that is, superconductor reverting to the normal state, the copper matrix carries the current without getting excessively hot. The superconductor is cooled below its T_c and carries the electric current again. This is the so-called cryogenic stability or cryostabilization design concept; namely, the superconductor is embedded in a large volume of low-resistivity copper and liquid helium is in intimate contact with all windings.

9.3.2 Ductile Alloys: Niobium–Titanium Alloys

Niobium–titanium alloys provide a good combination of superconducting and mechanical properties. A range of compositions is available commercially: Nb–44% Ti in the United Kingdom, Nb–46.5% Ti in the United States, and Nb–50% Ti in West Germany.

   In all these alloys, a $J_c > 1000$ A mm^{-2} at 4.2 K and an applied field of 7 T can be obtained by a suitable combination of mechanical working and annealing treatments. Strong flux pinning and therefore high J_c are obtained in these alloys by means of dislocation cell walls and precipitates. The flux pinning by precipitates becomes important in high-Ti alloys because the Nb-Ti phase diagram indicates precipitation of α-Ti in these alloys.

   As pointed out above, the condition of stability against flux motion requires that the superconductor be manufactured in the form of a composite system: extremely fine superconducting filaments embedded in a copper matrix provide flux stability and reduced losses caused by varying magnetic fields. Fortunately, Nb-Ti and

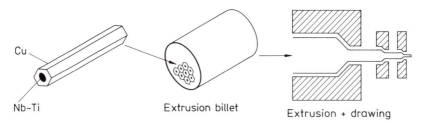

Cu

Nb-Ti   Extrusion billet
                                                              Extrusion + drawing

Fig. 9.4. Fabrication route for Nb-Ti/Cu composite superconductors.

copper are compatible and amenable to making filamentary composites. Figure 9.4 shows the essential steps in the fabrication of Nb-Ti composite superconductors. Annealed Nb-Ti rods are inserted into hexagonal shaped high-purity copper tubes. These rods are next loaded into an extrusion billet of copper, evacuated, sealed, and extruded. The extruded rods are cold drawn to an intermediate size and annealed to provide the necessary dislocation cell walls and precipitates for the flux pinning. This is followed by more cold drawing passes to the proper final size and a final anneal to get back the high conductivity of the copper matrix. Consider the specific case of Nb-50% Ti alloy. Its initial microstructure consists of a β solid solution. The necessary cell structure and dislocation density for flux pinning purposes will depend on the purity of the alloy and the size and distribution of the α particles after precipitation heat treatment. The α phase is nonsuperconducting. Its main function is to aid dislocation cell structure formation and the two together (α particles and dislocations) are responsible for flux pinning and thus contribute to high J_c values. The amount and distribution of α particles depend on the alloy chemical composition, processing, and annealing temperature and time. For a Nb-50% Ti alloy, 48 h at 375°C is generally used and results in about 11% of α particles. A greater amount of α will reduce the ductility of the alloy. Higher annealing temperatures result in excessive softening, fewer dislocations, and lower J_c. The precipitation treatment is generally followed by some cold working to refine the structure and obtain a high J_c. The precise amount of strain given in this last step is a function of the superconductor design, that is, the distribution of Nb-Ti filaments and the ratio of Nb-Ti/Cu in the cross section.

Figure 9.5a shows a compacted strand cable type superconductor made from 15 fine multifilamentary strands (strand diameter = 2.3 mm). Each strand consists of 1060 filaments of Nb-Ti (diameter = 50 μm) embedded in a copper matrix. A magnified picture of one of the strands is shown in Fig. 9.5b, while a scanning electron microscope picture is shown in Fig. 9.6. The Cu-Ni layers seen in Fig. 9.6 provide the low-conductivity barriers to prevent eddy current losses in alternating fields.

Niobium–titanium superconductors are finding an increasing number of applications. Commerically manufactured NMR (nuclear magnetic resonance) spectrometer systems, also called magnetic resonance imaging (MRI) systems, for medical diagnostics became available in the 1980s. Figure 9.7 shows one such system. The superconducting solenoid, made from Nb-Ti/Cu, is immersed in a cryogenic Dewar that allows the magnet system to operate about 3 months per refill, consuming about 4 ml of liquid helium per hour. Basically, the patient is placed in

Fig. 9.5. a Compacted strands cable type superconductor made from 15 multifilamentary strands (diameter = 2.3 mm). (Courtesy of Hitachi Cable Co.) b Magnified picture of one of the strands of part (a) containing 1060 filaments (diameter = 50 μm). (Courtesy of Hitachi Cable Co.)

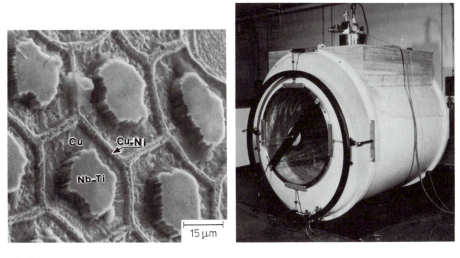

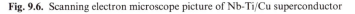

Fig. 9.6   Fig. 9.7

Fig. 9.6. Scanning electron microscope picture of Nb-Ti/Cu superconductor

Fig. 9.7. A magnetic resonance imaging system for clinical diagonstics. (Courtesy of Oxford-Airco.)

the bore of a very powerful superconducting magnet and high-frequency radio signals are used to probe the desired area. Atoms of specific elements in the body become excited and resonate in a characteristic way. A computer is used to construct the image of the organ from the resonance patterns. The great advantage of magnetic resonance imaging in clinical diagnostics is that it does not expose the patient to ionizing radiation and its possible harmful side effects. Of course, MRI techniques do not have to employ superconducting magnets but certain advantages exist with superconductors, for example, better homogeneity and resolution and higher field strengths than are available with conventional magnets. The disadvantage is that higher fields with superconductors lead to greater shielding problems. Another rather large-scale application of Nb-Ti/Cu superconducting magnets is in the magnetically levitated train. Japan National Railways has already tested such trains over a small stretch at speeds over 500 km h^{-1}. Nb-Ti/Cu superconductor composites are also employed in pulsed magnets for particle accelerators in high-energy physics and in magnetohydrodynamics.

Researchers at the General Electric Research and Development Center in Schenectady, New York, marked a major advance in cryogenics research in 1983 when they first successfully tested at *full load* an advanced superconducting electric generator for utility applications. The GE superconducting generator produced 20,600 kilovoltamperes of electricity, twice as much as could be produced by a conventional generator of a comparable size. The superconducting component of the generator was a 4-m-long rotor operating at liquid helium temperature, 4.2 K. The field windings were made from a Nb-Ti/Cu composite. Employing super-conducting materials, the designers made a generator that could develop a much stronger magnetic field than a conventional generator, permitting a significant reduction in the size of the generator for the same power output. Superconducting magnets do require cryogenic temperatures to operate, but the cost of this refrigera-tion is more than taken care of by the energy savings. One has only to remember that, in a superconducting machine working with almost zero resistance, the normal losses associated with the flow of electricity in rotor windings of a conventional machine are absent, resulting in a higher efficiency and reduced operating costs. Westinghouse has also developed similar superconducting generators.

One big problem in this development at GE was to prevent rotor windings movement under the intense centrifugal and magnetic forces exerted on them. The rotor spins at a speed at a 360 rpm. Thus, even an infinitesimally small movement of these components would generate enough heat by friction to quench the super-conductors. The GE researchers used a special vacuum epoxy-impregnation process to bond the Nb-Ti superconductors into *rock solid modules* and strong aluminum supports to hold the windings rigidly.

9.3.3 A-15 Superconductors: Nb$_3$Sn

For applications involving fields greater than 12 T and temperatures higher than 4.2 K, the ordered intermetallic compounds having an A-15 crystal structure are better suited than the Nb-Ti type. Nb$_3$Sn has a T_c of 18 K. It is easy to appreciate that the higher the T_c, the lower will be the refrigeration costs. Hence, the tremen-dous interest in high T_c oxide superconductors.

Superconducting magnets form an integral part of any thermonuclear fusion reactor for producing plasma-confining magnetic fields. It is estimated that these superconducting magnets will represent a sizable fraction of the capital cost of a fusion power plant. Their performance will affect the plasma as well as the plasma density — hence the importance of materials research aimed at improving the performance limits of these superconducting magnets. The main difference between the magnets used in a fusion reactor and in power transmission is that the former use superconductors at very high magnetic fields while the latter use them at low fields.

Nb$_3$Sn is the most widely used superconductor for high fields and high tempera-tures. A characteristic feature of these intermetallic compounds is their extreme brittleness (typically, a strain to fracture of 0.2% with no plasticity). Compare this to Nb-Ti which can be cold worked to a reduction in area approaching 100%.

Initially, the compound Nb_3Sn was made in the form of wires or ribbons either by diffusion of tin into niobium substrate in the form of a ribbon or by chemical vapor deposition. V_3Ga on a vanadium ribbon was also produced in this manner. The main disadvantages of these ribbon type superconductors were: (a) flux instabilities due to one wide dimension in the ribbon geometry and (b) limited flexibility in the ribbon width direction. Later on, with the realization that flux stability could be obtained by the superconductors in the form of extremely fine filaments, the filamentary composite approach to A-15 superconductor fabrication was adopted.

This breakthrough involving the composite route came through early in the 1970s. Tachikawa [2] showed that V_3Ga could be produced on vanadium filaments in a Cu-Ga matrix while Kaufmann and Pickett [3] demonstrated that Nb_3Sn could be obtained on niobium filaments from a bronze (Cu-Sn) matrix.

Figure 9.8 shows schematically the process of producing Nb_3Sn (or V_3Ga). Niobium rods are inserted into a bronze (Cu–13 wt.% Sn) extrusion billet. This billet is sealed and extruded into rods. This is called the first-stage extrusion. These first-stage extrusion rods are loaded into a copper can that has a tantalum or niobium barrier layer and extruded again. The second-stage extruded composite is cold drawn and formed into a cable, clad with more copper and compacted to form a monolithic conductor. This is finally given a heat treatment to convert the very fine niobium (vanadium) filaments into Nb_3Sn (V_3Ga). Specifically, for the Nb_3Sn superconductors, the ratio of cross-sectional areas of bronze matrix to that of niobium, R = Cu-Sn/Nb, is an important design parameter that strongly affects the J_c value. One must assure that the right amount of tin is available to form the Nb_3Sn phase. Too much of the bronze matrix will reduce the J_c value. Thus, one uses a

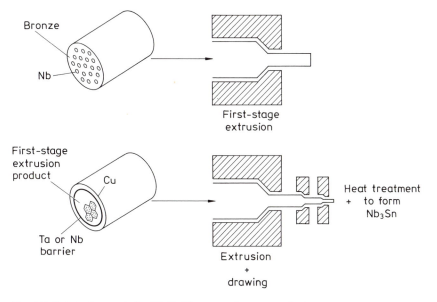

Fig. 9.8. Fabrication route for Nb_3Sn/Cu composite superconductor

Fig. 9.9. a Nb_3Sn/Cu composite superconductor used in high-field magnets. Each little dot in this picture is a 4-μm diameter Nb_3Sn filament. Hexagonal strips are niobium diffusion barriers for tin. (Courtesy of Hitachi Cable Co.) **b** Scanning electron microscope picture of an niobium/bronze composite before the heat treatment to form Nb_3Sn

bronze with about 13 wt.% tin, the limit of solubility of tin in copper, which yields enough tin without affecting the formability. Arriving at an optimum value of the ratio R is difficult because the resultant Nb_3Sn may not be stoichiometric. Superconductors meant for high fields (~ 15 T) should have a stoichiometric Nb_3Sn. One ensures this by using a high ratio of R (> 3) and sufficiently long times.

An example of a Nb_3Sn monolithic superconductor used in high-field magnets is shown in Fig. 9.9. Each little dot in this photograph (Fig. 9.9a) is a 4-μm Nb_3Sn filament. Wide strips in the hexagonal form are niobium and they serve as barriers to tin diffusion from the bronze into the copper. Tantalum is also used as a diffusion barrier. Figure 9.9b shows a scanning electron micrograph of an niobium/bornze composite before heat treatment tò form Nb_3Sn.

It has been shown [4, 5] that the rather good flux pinning and the consequent high critical currents in these superconductors are due to the grain boundaries. At the heat treatment temperature of about 650–700°C, the Nb_3Sn reaction layer formed consists of very fine grains, less than 80 nm (0.08 μm) in diameter. The critical current density for Nb_3Sn under these conditions is more than 2000 A mm^{-2} at 10 T and 4.2 K.

There exists a serious problem, something very much inherent to composites, with these A-15 type composite superconductors. This has to do with the thermal stresses that result from the differential expansion or contraction between the composite components [6, 7]. The different coefficients of thermal expansion of bronze and Nb_3Sn lead to rather large compressive strains in the Nb_3Sn layer when cooled from the reaction temperature (~ 1000 K) to 4.2 K. Luhman and Suenaga [8] showed that the T_c of Nb_3Sn varied with the strain applied to the Nb_3Sn layer by the bronze matrix. Later on, measurements of critical currents as a function of applied strain confirmed these results and one could explain the strain–critical current behavior in terms of the effects of strain on T_c and H_{c_2}. This understanding of strain–critical current behavior is used to good effect in the design of Nb_3Sn superconducting magnets. As pointed out, Nb_3Sn filaments degrade at a tensile strain of about 0.2%. Thus, if the superconductor is designed so that the Nb_3Sn filaments in the bronze matrix experience compression between 0.4 and 0.6%, then one can expect these composites to withstand applied strains of between 0.6 and 0.8% before fracture ensues in the Nb_3Sn filaments.

9.4 In Situ and Powder Metallurgy Fabrication [9]

Although the bronze route of manufacturing the superconductor has entered the phase of commercial production, there are some disadvantages involved. One must have frequent interruptions for annealing the work hardened bronze matrix. To shorten this process of fabrication, a number of in situ and powder metallurgy techniques are being tried. These in situ techniques involve melting of copper-rich Cu-Nb mixtures (~ 1800–$1850°C$), homogenization, and casting. Niobium is practically insoluble in copper at ambient temperature. Thus, the casting consists of niobium precipitates in a copper matrix. When this is cold drawn into wire, niobium is converted into fine filaments in the copper matrix. Tin is then plated onto the wire and diffused to form Nb_3Sn. The different melting and cooling techniques used include chill casting, continuous casting, levitation melting, and consumable electrode arc melting.

The powder metallurgy techniques involve mixing of copper and niobium powder, pressing, hot or cold extrusion, and drawing to a fine wire whch is coated or reacted with tin. The contamination of niobium with oxygen causes embrittlement and prevents its conversion into fine filaments. This has led to the addition of a third element (Al, Zr, Mg, or Ca) to the Cu-Nb mixture. This third element preferentially binds oxygen and leaves the niobium ductile.

References

1   J.E. Kunzler, E. Bachler, F.S.L. Hsu, and J.E. Wernick, *Phys. Rev. Lett.*, **6**, 89 (1961).
2   K. Tachikawa, in *Proceedings of the 3rd ICEC*, Illife Science and Technology Publishing, Surrey, U.K., 1970.
3   A.R. Kaufmann and J.J. Pickett, *Bull. Am. Phys. Soc.*, **15**, 833 (1970).
4   E. Nembach and K. Tachikawa, *J. Less Common Met.*, **19**, 1962 (1979).
5   R.M. Scanlan, W.A. Fietz, and E.F. Koch, *J. Appl. Phys.*, **46**, 2244 (1975).
6   K.K. Chawla, *Metallography*, **6**, 55 (1973).
7   K.K. Chawla, *Philos. Mag.*, **28**, 401 (1973).
8   T. Luhman and M. Suenaga, *Appl. Phys. Lett.*, **29**, 1 (1976).
9   R. Roberge and S. Foner, in *Filamentary A15 Superconductors*, Plenum Press, New York, 1980, p. 241.

Suggested Reading

E.W. Collings, *Design and Fabrication of Conventional and Unconventional Superconductors*, Noyes Publications, Park Ridge, NJ, 1984.

J.W. Ekin, in *Materials at Low Temperatures*, ASM, Metals Park, OH, 1983, p. 465.

S. Foner and B.B. Schwartz (eds.), *Superconducting Materials Science*, Plenum Press, New York, 1981.

T. Luhman and D. Dew-Hughes (eds.), *Metallurgy of Superconducting Materials*, Treatise on Material Science and Technology, vol. 14, Academic Press, New York, 1979.

R.P. Reed and A.F. Clark (eds.), *Advances in Cryogenic Engineering Materials*, vol. 28, Plenum Press, New York, 1982.

R.M. Scanlan, *Ann. Rev. Mater. Sci.*, **10**, 113 (1980).

M. Suenaga and A.F. Clark (eds.), *Filamentary A15 Superconductors*, Plenum Press, New York, 1980.

Part III

10. Micromechanics of Composites

In this chapter we consider the results of incorporating fibers in a matrix. The matrix, besides holding the fibers together, has the important function of transferring the applied load to the fibers. It is of great importance to be able to predict the properties of a composite, given the component properties and their geometric arrangement. We examine various micromechanical aspects of fibrous composites. A particularly simple case is the *rule of mixtures*, a rough tool that considers the composite properties as volume-weighted averages of the component properties. It is important to realize that the rule of mixtures works in only certain simple situations. Composite density is an example where the rule of mixtures is applied readily. In the case of mechanical properties, there are certain restrictions to its applicability. If more precise information is desired, it is better to use more sophisticated approaches based on the theory of elasticity.

10.1 Density

Consider a composite of mass m_c and volume v_c. The total mass of the composite is the sum total of the masses of fiber and matrix, that is,

$$m_c = m_f + m_m \tag{10.1}$$

The subscripts c, f, and m indicate composite, fiber, and matrix, respectively. Note that Eq. (10.1) is valid even in the presence of any voids in the composite. The volume of the composite, however, must include the volume of voids v_v. Thus,

$$v_c = v_f + v_m + v_v \tag{10.2}$$

Dividing Eq. (10.1) by m_c and Eq. (10.2) by v_c and denoting the mass and volume fractions by M_f, M_m and V_f, V_m, V_v, respectively, we can write

$$M_f + M_m = 1 \tag{10.3}$$

and

$$V_f + V_m + V_v = 1 \tag{10.4}$$

The composite density $\rho_c \ (= m/v)$ is given by

$$\rho_c = \frac{m_c}{v_c} = \frac{m_f + m_m}{v_c} = \frac{\rho_f v_f + \rho_m v_m}{v_c}$$

or

$$\rho_c = \rho_f V_f + \rho_m V_m \tag{10.5}$$

We can also derive an expression for ρ_c in terms of mass fractions. Thus,

$$\rho_c = \frac{m_c}{v_c} = \frac{m_c}{v_f + v_m + v_v} = \frac{m_c}{m_f/\rho_f + m_m/\rho_m + v_v}$$

$$= \frac{1}{M_f/\rho_f + M_m/\rho_m + v_v/m_c}$$

$$= \frac{1}{M_f/\rho_f + M_m/\rho_m + v_v/\rho_c v_c}$$

$$= \frac{1}{M_f/\rho_f + M_m/\rho_m + V_v/\rho_c} \tag{10.6}$$

We can use Eq. (10.6) to measure indirectly the volume fraction of voids in a composite. Rewriting Eq. (10.6), we obtain

$$\rho_c = \frac{\rho_c}{\rho_c[M_f/\rho_f + M_m/\rho_m] + V_v}$$

or

$$V_v = 1 - \rho_c \left(\frac{M_f}{\rho_f} + \frac{M_m}{\rho_m} \right) \tag{10.7}$$

10.2 Mechanical Properties

In this section we first describe some of the methods for predicting elastic constants, thermal properties, and transverse stresses in fibrous composites and then treat the mechanics of load transfer.

10.2.1 Prediction of Elastic Constants

Consider a unidirectional composite such as the one shown in Fig. 10.1. Assume that plane sections of this composite remain plane after deformation. Let us apply a force P_c in the fiber direction. Now, if the two components adhere perfectly and if they have the same Poisson ratio, then each component will undergo the same longitudinal elongation Δl. Thus, we can write for the strain in each component

$$\varepsilon_f = \varepsilon_m = \varepsilon_{cl} = \frac{\Delta l}{l} \tag{10.8}$$

where ε_{cl} is the strain in the composite in the longitudinal direction. This is called the *isostrain* or *action in parallel* situation. It was first treated by Voigt [1]. If both

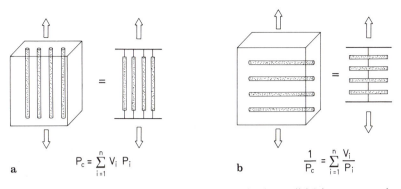

$$P_c = \sum_{i=1}^{n} V_i \, P_i$$

a

$$\frac{1}{P_c} = \sum_{i=1}^{n} \frac{V_i}{P_i}$$

b

Fig. 10.1. Unidirectional composite: **a** isostrain or action in parallel, **b** isostress or action in series

fiber and matrix are elastic, we can relate the stress σ in the two components to the strain ε_l by Young's modulus E. Thus,

$$\sigma_f = E_f \varepsilon_{cl} \quad \text{and} \quad \sigma_m = E_m \varepsilon_{cl}$$

Let A_c be the cross-sectional area of the composite, A_m that of the matrix, and A_f that of all the fibers. Then, from the equilibrium of forces in the fiber direction, we can write

$$P_c = P_f + P_m$$

or

$$\sigma_{cl} A_c = \sigma_f A_f + \sigma_m A_m \tag{10.9}$$

From Eqs. (10.8) and (10.9), we get

$$\sigma_{cl} A_c = (E_f A_f + E_m A_m)\varepsilon_{cl}$$

or

$$E_{cl} = \frac{\sigma_{cl}}{\varepsilon_{cl}} = E_f \frac{A_f}{A_c} + E_m \frac{A_m}{A_c}$$

Now, for a given composite length, $A_f/A_c = V_f$ and $A_m/A_c = V_m$. Then the above expression can be simplified to

$$E_{cl} = E_f V_f + E_m V_m = E_{11} \tag{10.10}$$

Equation (10.10) is called the rule of mixtures for Young's modulus in the fiber direction. A similar expression can be obtained for the composite longitudinal strength from Eq. (10.9), namely,

$$\sigma_{cl} = \sigma_f V_f + \sigma_m V_m \tag{10.11}$$

For properties in the transverse direction, we can represent the simple unidirectional composite by what is called the *action-in-series* or *isostress* situation; see Fig. 10.1b. In this case, we group the fibers together as a continuous phase normal to the stress. Thus, we have equal stresses in the two components and the model is equivalent to that treated by Reuss [2]. For loading transverse to the fiber direction, we have

$$\sigma_{ct} = \sigma_f = \sigma_m$$

while the total displacement of the composite in the thickness direction, t_c, is the sum of displacements of the components, that is,

$$\Delta t_c = \Delta t_m + \Delta t_f$$

Dividing throughout by t_c, the composite gage length, we obtain

$$\frac{\Delta t_c}{t_c} = \frac{\Delta t_m}{t_c} + \frac{\Delta t_f}{t_c}$$

Now $\Delta t_c/t_c = \varepsilon_{ct}$, the composite strain in the transverse direction, while Δt_m and Δt_f equal the strains in the matrix and fiber times their respective gage lengths; that is, $\Delta t_m = \varepsilon_m t_m$ and $\Delta t_f = \varepsilon_f t_f$. Then

$$\varepsilon_{ct} = \frac{\Delta t_c}{t_c} = \frac{\Delta t_m}{t_m}\frac{t_m}{t_c} + \frac{\Delta t_f}{t_f}\frac{t_m}{t_c}$$

or

$$\varepsilon_{ct} = \varepsilon_m \frac{t_m}{t_c} + \varepsilon_f \frac{t_f}{t_c} \tag{10.12}$$

For a given composite cross-sectional area under the applied load, the volume fractions of fiber and matrix can be written

$$V_m = \frac{t_m}{t_c} \quad \text{and} \quad V_f = \frac{t_f}{t_c}$$

This simplifies Eq. (10.12) to

$$\varepsilon_{ct} = \varepsilon_m V_m + \varepsilon_f V_f \tag{10.13}$$

Considering both components to be in the elastic regime and remembering that $\sigma_{ct} = \sigma_f = \sigma_m$ in this case, we can write Eq. (10.13) as

$$\frac{\sigma_{ct}}{E_{ct}} = \frac{\sigma_{ct}}{E_m} V_m + \frac{\sigma_{ct}}{E_f} V_f$$

or

$$\frac{1}{E_{ct}} = \frac{V_m}{E_m} + \frac{V_f}{E_f} = \frac{1}{E_{22}} \tag{10.14}$$

Relationships given by Eqs. (10.5), (10.10), (10.11), (10.13), and (10.14) are commonly referred to as rules of mixtures. Figure 10.2 shows the plots of Eqs. (10.10) and (10.14). The reader should appreciate that these relationships and their variants are but rules of thumb obtained from a simple strength of materials approach. More comprehensive micromechanical models, based on the theory of elasticity, can be and should be used to obtain the elastic constants of fibrous composites. We describe below, albeit very briefly, some of these. A critique of these models has been made by Chamis and Sendeckyj [3].

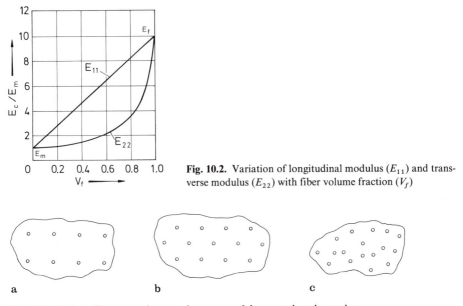

Fig. 10.2. Variation of longitudinal modulus (E_{11}) and transverse modulus (E_{22}) with fiber volume fraction (V_f)

Fig. 10.3. Various fiber arrays in a matrix: **a** square, **b** hexagonal, and **c** random

10.2.2 Micromechanical Approach

An anisotropic body in the most general case has 21 independent elastic constants [4]. An isotropic body, on the other hand, has only two independent elastic constants. In such a body, when a tensile stress is applied in the z direction, a tensile strain e results in that direction. In addition to this, because of the Poisson ratio effect, two equal compressive strains ($\varepsilon_x = \varepsilon_y$) result in the x and y directions. In a generally anisotropic body, the two transverse strain components are not equal. In fact, as we shall see in Chap. 11, in such a body, tensile loading can result in tensile as well as shear strains. The large number of independent elastic constants (21 in the most general case, i.e., no symmetry elements) represents the complexity of the situation. Any symmetry elements present will reduce the number of independent elastic constants [4].

A composite containing uniaxially aligned fibers will have a plane of symmetry perpendicular to the fiber direction (i.e., material on one side of the plane will be the mirror image of the material on the other side). Such a material will have 13 independent elastic constants. Additional symmetry elements, depending on the fiber arrangement, can be present. Figure 10.3 shows square, hexagonal, and random fiber arrays in a matrix. A square array of fibers, for example, will have symmetry planes parallel to the fibers as well as perpendicular to them. Such a material is an orthotropic material (three mutually perpendicular planes of symmetry) and possesses nine independent elastic constants [4]. Hexagonal and random arrays of aligned fibers are transversely isotropic and have five independent elastic constants. These five constants as well as the stress–strain relationships, as derived by Hashin and Rosen [5, 6], are given in Table 10.1. There are two Poisson

Table 10.1. Elastic moduli of a transversely isotropic fibrous composite

$E = C_{11} - \dfrac{2C_{12}^2}{C_{22} + C_{23}}$	$K_{23} = \frac{1}{2}(C_{22} + C_{23})$
$G = G_{12} = G_{13} = C_{44}$	$G_{23} = \frac{1}{2}(C_{22} - C_{23})$
$\nu = \nu_{13} = \nu_{31} = \dfrac{1}{2}\left(\dfrac{C_{11} - E}{K_{23}}\right)^{1/2}$	$\nu_{23} = \dfrac{K_{23} - \phi G_{23}}{K_{23} + \phi G_{23}}$
$E_2 = E_3 = \dfrac{4G_{23}K_{23}}{K_{23} + \phi G_{23}}$	$\phi = 1 + \dfrac{4K_{23}\nu^2}{E}$

Stress–Strain Relationships

$\varepsilon_{11} = \dfrac{1}{E_1}[\sigma_{11} - \nu(\sigma_{22} + \sigma_{33})]$	$\varepsilon_{22} = \varepsilon_{33} = \dfrac{1}{E_2}(\sigma_{22} - \nu\sigma_{33}) - \dfrac{\nu}{E}\sigma_{11}$
$\gamma_{12} = \gamma_{13} = \dfrac{1}{G}\sigma_{12}$	$\gamma_{23} = \dfrac{2(1 + \nu_{23})}{E_2}\sigma_{23}$

Source: Adapted with permission from Ref. 5.

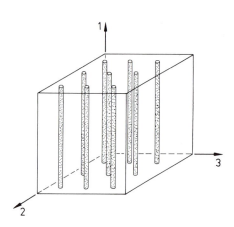

Fig. 10.4. A transversely isotropic fiber composite: plane transverse to fibers (2-3 plane) is isotropic

ratios: one gives the transverse strain caused by an axially applied stress and the other gives the axial strain caused by a transversely applied stress. The two are not independent but are related. Thus, the number of independent elastic constants for a transversely isotropic composite is five. Note that the total number of independent elastic constants in Table 10.1 is five (count the number of Cs).

Chamis [7] has summarized the elastic constants for a transversely isotropic composite in terms of the elastic constants of the two components. He gives the relationships for a thin sheet or ply, but they are more general and valid for any transversely isotropic composite such as the one shown in Fig. 10.4. Since the plane 2-3 is isotropic in Fig. 10.4, the properties in directions 2 and 3 are identical. Chamis treats the matrix as an isotropic material, while the fiber is treated as an anisotropic material. Thus, E_m and ν_m are the two constants required for the matrix while five constants (E_{f1}, E_{f2}, G_{f12}, G_{f23}, and ν_{f12}) are required for the fiber. Table 10.2

Table 10.2. Elastic constants of a transversely isotropic composite in terms of component constants (matrix isotropic, fiber anisotropic)

Longitudinal modulus	$E_{11} = E_{f1}V_f + E_mV_m$
Transverse modulus	$E_{22} = E_{33} = \dfrac{E_m}{1 - \sqrt{V_f}(1 - E_m/E_{f2})}$
Shear modulus	$G_{12} = G_{13} = \dfrac{G_m}{1 - \sqrt{V_f}(1 - G_m/G_{f12})}$
Shear modulus	$G_{23} = \dfrac{G_m}{1 - \sqrt{V_f}(1 - G_m/G_{f23})}$
Poisson ratio	$\nu_{12} = \nu_{13} = \nu_{f12}V_f + \nu_mV_m$
Poisson ratio	$\nu_{23} = \dfrac{E_{22}}{2G_{23}} - 1$

Source: Adapted with permission from Ref. 7.

gives these equations. The five independent constants are E_{11}, E_{22}, G_{12}, G_{13} and ν_{12}.

Frequently, composite structures are fabricated by stacking up thin sheets of unidirectional composites called plies in an appropriate orientation sequence dictated by elasticity theory (see Chap. 11). It is of interest to know the properties, such as the elastic constants, and the strength characteristics of a ply. In particular, it is of value if we are able to predict the lamina characteristics starting from the individual component characteristics. Later in the macromechanical analysis we treat a ply as a homogeneous but thin orthotropic material. Elastic constants in the thickness direction can be ignored in such a ply, leaving four independent elastic constants, namely, E_{11}, E_{22}, ν_{12}, and G_{12}, one less than the number for a thick but transversely isotropic material. The missing constant is G_{23}, the transverse shear modulus in the 2-3 plane normal to the fiber axis.

A brief description of the various micromechanical techniques used for predicting the elastic constants is given below. Chamis and Sendeckyj [3] have presented a critique of these methods. Following this brief description we give an account of a set of empirical equations, called Halpin–Tsai equations, that has been developed for predicting the elastic constants of a fiber composite.

In the so-called self-consistent field methods [3], approximations of phase geometries are made and a simple representation of the response field is obtained. The phase geometry is represented by one single fiber embedded in a matrix cylinder. This outer cylinder is embedded in an unbounded homogeneous material whose properties are taken to be equivalent to those of average composite properties. The matrix under a uniform load at infinity introduces a uniform strain field in the fiber. Elastic constants are obtained from this strain field. The results obtained are independent of fiber arrangements in the matrix and are reliable at low fiber volume fractions (V_f), reasonable at intermediate V_f, and unreliable at high V_f [8]. Exact methods deal with specific geometries, for example, fibers arranged in a hexagonal, square, or rectangular array in a matrix. The elasticity problem is then solved by a series development, a complex variable technique, or a numerical solution.

The variational or bounding methods focus on the upper and lower bounds on elastic constants. They do not predict properties directly. If, however, the upper and lower bounds coincide, then the property is determined exactly. Frequently, the upper and lower bounds are well separated. When these bounds are close enough, we can safely use them as indicators of the material behavior. It turns out that this is the case for longitudinal properties of a unidirectional lamina. Hill [9] derived bounds for the ply elastic constants that are analogous to those derived by Hashin and Rosen [5, 6]. In particular, Hill put rigorous bounds on the longitudinal Young's modulus, E_{11}, in terms of the bulk modulus in plane strain (k_p), Poisson's ratio (v), and the shear modulus (G) of the two phases. No restrictions were made on the fiber form or packing geometry. The term k_p is the modulus for lateral dilatation with zero longitudinal strain and is given by

$$k_p = \frac{E}{2(1 - 2v)(1 + v)}$$

The bounds on the longitudinal modulus, E_{11}, are

$$\frac{4V_f V_m (v_f - v_m)^2}{(V_f/k_{pm}) + (V_m/k_{pf}) + 1/G_m} \leqslant E_{11} - E_f V_f - E_m V_m$$

$$\leqslant \frac{4V_f V_m (v_f - v_m)^2}{(V_f/k_{pm}) + (V_m/k_{pf}) + 1/G_f} \tag{10.15}$$

Equation (10.15) shows that the deviations from the rule of mixtures [Eq. (10.10)] are quite small ($<2\%$). We may verify this by substituting some values of practical composites such as carbon or boron fibers in an epoxy matrix or a metal matrix composite such as tungsten in a copper matrix. Note that the deviation from the rule of mixtures value comes from the $(v_m - v_f)^2$ factor. For $v_f = v_m$, we have E_{11} given exactly by the rule of mixtures.

Hill [9] also showed that for a unidirectionally aligned fiber composite

$$v_{12} \gtrless v_f V_f + v_m V_m \quad \text{accordingly as} \quad (v_f - v_m)(k_{pf} - k_{pm}) \gtrless 0 \tag{10.16}$$

Generally, $v_f < v_m$ and $E_f \gg E_m$. Then, v_{12} will be less than that predicted by the rule of mixtures ($= v_f V_f + v_m V_m$). It is easy to see that the bounds on v_{12} are not as close as the ones on E_{11}. This is because $v_f - v_m$ appears in the case of v_{12} [Eq. (10.16)] while $(v_f - v_m)^2$ appears in the case of E_{11} [Eq. (10.15)]. In the case where $v_f - v_m$ is very small, the bounds are close enough to allow us to write

$$v_{12} \simeq v_f V_f + v_m V_m \tag{10.17}$$

10.2.3 Halpin–Tsai Equations

Halpin, Tsai, and Kardos [10–12] have empirically developed some generalized equations that readily give quite satisfactory results compared to the complicated micromechanical equations. These equations are quite accurate at low fiber volume fractions. They are also useful in determining the properties of composites that contain discontinuous fibers oriented in the loading directions. One writes a single equation of the form

$$\frac{p}{p_m} = \frac{1 + \xi\eta V_f}{1 - \eta V_f} \tag{10.18}$$

$$\eta = \frac{p_f/p_m - 1}{p_f/p_m + \xi} \tag{10.19}$$

where p represents composite moduli, for example, E_{11}, E_{22}, G_{12} or G_{23}; p_m and p_f are the corresponding matrix and fiber moduli, respectively; V_f is the fiber volume fraction; and ξ is a measure of the reinforcement which depends on boundary conditions (fiber geometry, fiber distribution, and loading conditions). The term ξ is an empirical factor that is used to make Eq. (10.18) conform to the experimental data.

The function η in Eq. (10.19) is constructed in such a way that when $V_f = 0$, $p = p_m$ and when $V_f = 1, p = p_f$. Furthermore, the form of η is such that

$$\frac{1}{p} = \frac{V_m}{p_m} + \frac{V_f}{p_f} \quad \text{for } \xi \to 0$$

and

$$p = p_f V_f + p_m V_m \quad \text{for } \xi \to \infty$$

These two extremes (not necessarily tight) bound the composite properties. Thus, values of ξ between 0 and ∞ will give an expression for p between these extremes. Some typical values of ξ are given in Table 10.3. Thus, we can cast the Halpin–Tsai equations for the transverse modulus as

$$\frac{E_{22}}{E_m} = \frac{1 + \xi\eta V_f}{1 - \eta V_f} \quad \text{and} \quad \eta = \frac{E_f/E_m - 1}{E_f/E_m + \xi} \tag{10.20}$$

Comparing these expressions with exact elasticity solutions, one can obtain the value of ξ. Whitney [13] suggests $\xi = 1$ or 2 for E_{22}, depending on whether a hexagonal or square array of fibers is used.

Nielsen [14] has modified the Halpin–Tsai equations to include the maximum packing fraction ϕ_{max} of the reinforcement. His equations are

Table 10.3. Values of ξ for some uniaxial composites

Modulus	ξ
E_{11}	$2(l/d)$
E_{22}	0.5
G_{12}	1.0
G_{21}	0.5
K	0

Source: Adapted from Ref. 14, by courtesy of Marcel Dekker, Inc.

$$\frac{p}{p_m} = \frac{1 + \xi \eta V_f}{1 - \eta \Psi V_f}$$

$$\eta = \frac{p_f/p_m - 1}{p_f/p_m + \xi} \tag{10.21}$$

$$\Psi \simeq 1 + \left(\frac{1 - \phi_{max}}{\phi_{max}^2}\right) V_f$$

where ϕ_{max} is the maximum packing factor. It allows one to take into account the maximum packing fraction. For a square array of fibers, $\phi_{max} = 0.785$, while for a hexagonal arrangement of fibers, $\phi_{max} = 0.907$. In general, ϕ_{max} is between these two extremes and near the random packing, $\phi_{max} = 0.82$.

10.2.4 Transverse Stresses

When a fibrous composite consisting of components with different elastic moduli is uniaxially loaded, stresses in transverse directions arise because of the Poisson ratio differences between the matrix and fiber, that is, because the two components have different contractile tendencies. We follow here Kelly's [15] treatment of this important but, unfortunately, not well appreciated subject.

Consider a unit fiber reinforced composite consisting of a single fiber (radius a) surrounded by its shell of matrix (outer radius b) as shown in Fig. 10.5. The composite as a whole is thought to be built of an assembly of such unit composites, a reasonably valid assumption at moderate fiber volume fractions. We apply an axial load (direction z) to the composite. Owing to the obvious cylindrical symmetry, we treat the problem is polar coordinates, r, θ, and z. It follows from the axial symmetry that the stress and strain are independent of angle and are functions only of r, which simplifies the problem. We can write Hooke's law for this situation as

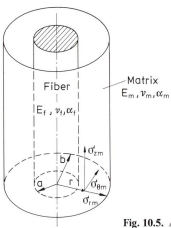

Fig. 10.5. A single fiber surrounded by its matrix shell

$$\begin{bmatrix} e_r & 0 & 0 \\ 0 & e_\theta & 0 \\ 0 & 0 & e_z \end{bmatrix} = \frac{1+v}{E} \begin{bmatrix} \sigma_r & 0 & 0 \\ 0 & \sigma_\theta & 0 \\ 0 & 0 & \sigma_z \end{bmatrix} - \frac{v}{E}(\sigma_r + \sigma_\theta + \sigma_z) \begin{bmatrix} 1 & 0 & 0 \\ 0 & 1 & 0 \\ 0 & 0 & 1 \end{bmatrix} \tag{10.22}$$

where e is the strain, σ is the stress, v is the Poisson ratio, E is Young's modulus in the longitudinal direction, and the subscripts r, θ, and z refer to the radial, circumferential, and axial directions, respectively. The only equilibrium equation for this problem is

$$\frac{d\sigma_r}{dr} + \frac{\sigma_r - \sigma_\theta}{r} = 0 \tag{10.23}$$

Also, for the plane strain condition, we can write for the strain components, in terms of displacements,

$$e_r = \frac{du_r}{dr} \qquad e_\theta = \frac{u_r}{r} \qquad e_z = \text{const} \tag{10.24}$$

where u_r is the radial displacement.

From Eq. (10.22) we have, after some algebraic manipulation,

$$\frac{\sigma_\theta}{K} = (1-v)e_\theta + v(e_r + e_z)$$
$$\frac{\sigma_r}{K} = (1-v)e_r + v(e_\theta + e_z) \tag{10.25}$$

where

$$K = \frac{E}{(1+v)(1-2v)}$$

From Eqs. (10.24) and (10.25), we get

$$\frac{\sigma_\theta}{K} = v\frac{du_r}{dr} + (1-v)\frac{u_r}{r} + ve_z$$
$$\frac{\sigma_r}{K} = (1-v)\frac{du_r}{dr} + v\frac{u_r}{r} + ve_z \tag{10.26}$$

Substituting Eq. (10.26) in Eq. (10.23), we obtain the following differential equation in terms of the radial displacement u_r:

$$\frac{d^2u_r}{dr^2} + \frac{1}{r}\frac{du_r}{dr} - \frac{u_r}{r^2} = 0 \tag{10.27}$$

Equation (10.27) is a common differential equation is elasticity problems with rotational symmetry [16] and its solution is

$$u_r = Cr + \frac{C'}{r} \tag{10.28}$$

where C and C' are constants of integration to be determined by using boundary conditions. Now, Eq. (10.28) is valid for displacements in both components, that is, fiber and matrix. Let us designate the central component by subscript 1 and the sleeve by subscript 2. Thus, we can write the displacements in the two components as

$$u_{r1} = C_1 r + \frac{C_2}{r}$$

(10.29)

$$u_{r2} = C_3 r + \frac{C_4}{r}$$

The boundary conditions can be expressed as follows:

1. At the free surface, the stress is zero, that is, $\sigma_{r2} = 0$ at $r = b$.
2. At the interface, the continuity condition requires that at $r = a$, $u_{r1} = u_{r2}$ and $\sigma_{r1} = \sigma_{r2}$.
3. The radial displacement must vanish along the symmetry axis, that is, at $r = 0$, $u_{r1} = 0$.

The last boundary condition immediately gives $C_2 = 0$, because otherwise u_{r1} will become infinite at $r = 0$. Applying the other boundary conditions to Eqs. (10.26) and (10.29), we obtain three equations with three unknowns. Knowing these integration constants, we obtain u and thus the stresses in the two components. It is convenient to develop an expression for radial pressure p at the interface. At the interface $r = a$, if we equate σ_{r2} to $-p$, then after some tedius manipulations it can be shown that

$$p = \frac{2e_z(v_2 - v_1)V_2}{V_1/k_{p2} + V_2/k_{p1} + 1/G_2}$$

(10.30)

where k_p is the plane strain bulk modulus equal to $E/2(1 + v)(1 - 2v)$. The expressions for the stresses in the components involving p are

Component 1:

$$\sigma_{r1} = \sigma_{\theta1} = -p$$

(10.31)

$$\sigma_{z1} = E_1 e_z - 2v_1 p$$

Component 2:

$$\sigma_{r2} = p\left(\frac{a^2}{b^2 - a^2}\right)\left(1 - \frac{b^2}{r^2}\right)$$

$$\sigma_{\theta2} = p\left(\frac{a^2}{b^2 - a^2}\right)\left(1 + \frac{b^2}{r^2}\right)$$

(10.32)

$$\sigma_{z2} = E_2 e_z + 2v_2 p\left(\frac{a^2}{b^2 - a^2}\right)$$

Note that p is positive when the central component 1 is under compression, that is, when $v_1 < v_2$.

Figure 10.6 shows the stress distribution schematically in a fiber composite (1 = fiber, 2 = matrix). We can draw some inferences from Eqs. (10.31) and (10.32) and Fig. 10.6.

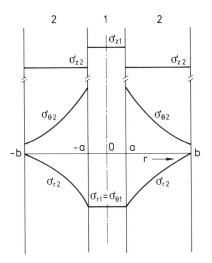

Fig. 10.6. Three-dimensional stress distribution (schematic) in the unit composite shown in Fig. 10.5. Transverse stresses (σ_r and σ_θ) result from the differences in the Poisson ratios of the fiber and matrix

1. Axial stress is uniform in components 1 and 2, although its magnitude is different in the two and depends on the respective elastic constants.
2. In the central component 1, σ_{r1} and $\sigma_{\theta 1}$ are equal in magnitude and sense. In the sleeve 2, σ_{r2} and $\sigma_{\theta 2}$ vary as $1 - b^2/r^2$ and $1 + b^2/r^2$, respectively.
3. When the Poisson ratio difference ($v_2 - v_1$) goes to zero, σ_r and σ_θ go to zero; that is, the rheological interaction will vanish.
4. Because of the relatively small difference in the Poisson ratios of the components of a metallic composite, the transverse stresses that develop in the elastic regime will be relatively small.

Kelly and Lilholt [17] invoked the existence of these transverse stresses owing to the Poisson ratio difference to explain the observed high strength and high work hardening rate of the copper matrix in the tungsten fiber/copper matrix composite compared to these same characteristics in unreinforced copper. The fact that the derived in situ copper stress–strain curve dropped after tungsten entered the plastic regime, thus eliminating the Poisson ratio difference, showed that the transverse stresses resulting from different contractile tendencies of the components do indeed alter the matrix state of stress in the composite.

10.3 Thermal Properties

Various equations have been proposed for obtaining the thermal expansion coefficients of a composite, knowing the material constants of the components [18–20]. Different equations predict very different values of expansion coefficients for a given composite. Almost all expressions, however, predict expansion coefficient values less than that given by a simple rule of mixtures ($= \alpha_f V_f + \alpha_m V_m$). This is because these equations take into account the important fact that the presence of fibers introduces mechanical constraint on the matrix. A fiber results in a greater constraint than a particle.

Kerner [19] developed the following expression for the volumetric expansion coefficient of a composite consisting of spherical particles dispersed in a matrix:

$$\alpha_c = \alpha_m V_m + \alpha_p V_p - (\alpha_m - \alpha_p) V_m V_p \frac{(1/K_m) - (1/K_p)}{(V_m/K_p) + (V_p/K_m) + (3/4G_m)} \tag{10.33}$$

where the subscripts m and p denote the matrix and the reinforcement, respectively; α_c, α_m, and α_p are the volumetric expansion coefficients of the composite, the matrix, and the reinforcement, respectively; and K_m and K_p are the bulk moduli of the two phases. Equation (10.33) does not differ significantly from the rule of mixtures because the particulate reinforcement constrains the matrix a lot less than fibers.

Unidirectionally aligned fiber composites have two (or sometimes three) thermal expansion coefficients: α_l in the longitudinal direction and α_t in the transverse direction. Fibers generally have a lower α than that of the matrix, and thus the former mechanically constrain the latter. This results in an α_{cl} smaller than α_{ct} for the composite. At low fiber volume fractions, it is not unusual to find α_{ct} greater than that of the matrix in isolation. The long stiff fibers prevent the matrix from expanding in the longitudinal direction and as a result the matrix is forced to expand more than usual in the transverse direction.

Schapery [20] has derived the following expression for the expansion coefficient of a fibrous composite, assuming that the Poisson ratios of the components are not very different. The longitudinal expansion coefficient for the composite is

$$\alpha_{cl} = \frac{\alpha_m E_m V_m + \alpha_f E_f V_f}{E_m V_m + E_f V_f} \tag{10.34}$$

and the transverse expansion coefficient is

$$\alpha_{ct} \simeq (1 + v_m)\alpha_m V_m + (1 + v_f)\alpha_f V_f - \alpha_{cl}\bar{v} \tag{10.35}$$

where α_{cl} is given by Eq. (10.34) and $\bar{v}$ is the approximate Poisson ratio of the composite given by $\bar{v} = v_f V_f + v_m V_m$. For high fiber volume fractions, that is, $V_f \gtrsim 0.2$ or 0.3, α_{ct} is given by

$$\alpha_{ct} \simeq (1 + v_m)\alpha_m V_m + \alpha_f V_f \tag{10.36}$$

Equations (10.34)–(10.36) are plotted in Fig. 10.7 for alumina fibers in an aluminum matrix. Note the marked anisotropy in the expansion for aligned fibrous composites. The expansion coefficient for a composite containing randomly oriented fibers in three dimensions is given by

$$\alpha \simeq \frac{\alpha_{cl} + 2\alpha_{ct}}{3}$$

where α_{cl} and α_{ct} are given by Eqs. (10.34) and (10.35), respectively.

Behrens [21] showed that the thermal conductivity (k) of a fiber composite in the fiber direction is given by

$$k_{cl} = k_f V_f + k_m V_m \tag{10.37}$$

Expressions for the general thermal properties of transversely isotropic fiber composites, as summarized by Chamis [7], are given in Table 10.4.

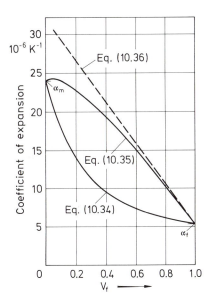

Fig. 10.7. Longitudinal and transverse linear thermal expansion coefficients versus fiber volume fraction for alumina fiber in an aluminum matrix

Table 10.4. Thermal properties of a transversely isotropic composite (matrix isotropic, fiber anisotropic)

Heat capacity	$C = \dfrac{1}{\rho}(V_f \rho_f C_f + V_m \rho_m C_m)$
Longitudinal conductivity	$k_{11} = V_f k_{f1} + V_m k_m$
Transverse conductivity	$k_{22} = k_{33} = (1 - \sqrt{V_f})k_m + \dfrac{k_m \sqrt{V_f}}{1 - \sqrt{V_f}(1 - k_m/k_{f2})}$
Longitudinal thermal expansion coefficient	$\alpha_{11} = \dfrac{V_f E_{f1}\alpha_{f1} + V_m E_m \alpha_m}{E_{f1} V_f + E_m V_m}$
Transverse thermal expansion coefficient	$\alpha_{22} = \alpha_{33} = \alpha_{f2}\sqrt{V_f} + \alpha_m(1 - \sqrt{V_f})\left(1 + \dfrac{V_f V_m E_{f1}}{E_{f1} V_f + E_m V_m}\right)$

Source: Adapted with permission from Ref. 7.

10.3.1 Hygrothermal Stresses

Hygroscopy, meaning water absorption, can be a problem with polymeric resins and natural fibers because it can lead to swelling. Swelling can lead to stresses if, as is likely in a composite, the material is not allowed to expand freely because of the presence of the matrix. Thermal stresses will result in a material when the material is not allowed to expand or constract freely because of a constraint. Consider the passage of a composite from a reference state, where the body is stress free and relaxed at a temperature T, concentration of moisture $C = 0$, and external stress $\sigma = 0$, to a final state where the body has hygrothermal as well as external

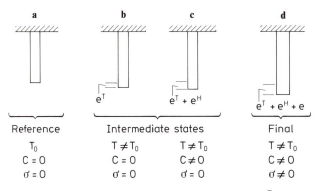

Fig. 10.8. a Strain-free reference state; **b** thermal strain (e^T); **c** hygral (e^H) and thermal (e^T) strains; **d** hygral (e^H), thermal (e^T), and mechanical strain; and (**e**) final state

stresses. We can consider that the final state of the composite with hygrothermal stresses and external loading is attained via two intermediate stages shown in Fig. 10.8. The final strain in the body can then be written

$$e_i = \underbrace{e_i^T + e_i^H}_{\text{nonmechanical strains}} + \underbrace{S_{ij}\sigma_j}_{\text{mechanical strain}} \qquad (10.38)$$

If we regard the composite as transversely isotropic, we can write

$$e_{xy}^T = e_{xy}^H = 0$$

$$e_y^T = e_z^T \quad \text{and} \quad e_y^H = e_z^H$$

$$e_i^T = \alpha_i(T - T_0) \qquad\qquad (10.39)$$

$$e_i^H = \beta_i C$$

where the α_i are thermal expansion coefficients (K^{-1}), the β_i are nondimensional swelling coefficients, T is the temperature, T_0 is the equilibrium temperature, C is the concentration of water vapor, and the superscripts H and T indicate hygral and thermal strain components, respectively.

Total volumetric hygrothermal strain can be expressed as the sum of the diagonal terms of the strain matrix. It is important to note that the thermal and hygral effects are dilatational only, i.e., they cause only expansion or contraction but do not affect the shear components. Thus,

$$\frac{\Delta V}{V} = e_x^T + e_y^T + e_z^T + e_x^H + e_y^H + e_z^H$$

$$= e_x^T + 2e_y^T + e_x^H + 2e_y^H \qquad\qquad (10.40)$$

Hygrothermal stresses are very important in polymer matrix composites (see also Chaps. 3 and 4).

During curing or solidification of the matrix around fibers, a high magnitude of shrinkage stresses can result. The interfacial pressure developed during curing is akin to that obtained upon embedding a cylinder of radius $r + \delta r$ in a cylindrical hole of radius r. Specifically, the thermal stresses generated depend on the fiber volume fraction, fiber geometry, thermal mismatch ($\Delta\alpha$), and the modulus ratio, E_f/E_m. Generally, $\alpha_m > \alpha_f$; that is, on cooling from T_0 to T ($T_0 > T$), the matrix would tend to contract more than the fibers, causing the fibers to experience axial compression. In extreme cases, fiber buckling can also lead to the generation of interfacial shear stresses. This problem of thermal stresses in composite materials is a most serious and important problem. It is worth repeating that thermal stresses are internal stresses that arise when there exists a constraint on the free dimensional change of a body [22]. In the absence of this constraint, the body can experience free thermal strains without any accompanying thermal stresses. The constraint can have its origin in (a) a temperature gradient, (b) crystal structure anisotropy (e.g., noncubic structure), (c) phase transformations resulting in a volume change, and (d) a composite material made of dissimilar materials (i.e., materials having different α's). The thermal gradient problem is a serious one in ceramic materials in general. A thermal gradient ΔT is inversely related to the thermal diffusivity a of a material. Thus,

$$\Delta T = \phi\left(\frac{1}{a}\right) = \phi\left(\frac{C_p\rho}{k}\right) \tag{10.41}$$

where C_p is the specific heat, ρ is the density, and k is the thermal conductivity. Metals generally have high thermal diffusivity and any thermal gradients that might develop are dissipated rather quickly. It should be emphasized that in composite materials even a uniform temperature change (i.e., no temperature gradient) will result in thermal stresses owing to the ever present thermal mismatch [22]. Thermal stresses resulting from a thermal mismatch will generally have an expression of the form

$$\sigma = f(E\,\Delta\alpha\,\Delta T) \tag{10.42}$$

We describe below the three-dimensional thermal stress state in a composite consisting of a central fiber surrounded by its shell of matrix; see Fig. 10.5.

The elasticity problem is basically the same as the one discussed for transverse stresses. We use polar coordinates, r, θ, and z (see Fig. 10.5). Axial symmetry makes shear stresses go to zero and the principal stresses are independent of θ. At low volume fractions, the outer cylinder is the matrix and the inner cylinder is the fiber. The expression for strain has an $\alpha\,\Delta T$ component. Thus, for component 2,

$$e_{r2} = \frac{\sigma_{r2}}{E_2} - \frac{v_2}{E_2}(\sigma_{\theta2} + \sigma_{z2}) + \alpha_2\,\Delta T$$

$$e_{\theta2} = \frac{\sigma_{\theta2}}{E_2} - \frac{v_2}{E_2}(\sigma_{r2} + \sigma_{z2}) + \alpha_2\,\Delta T$$

$$e_{z2} = \frac{\sigma_{z2}}{E_2} - \frac{v_2}{E_2}(\sigma_{r2} + \sigma_{\theta2}) + \alpha_2\,\Delta T$$

The resultant stresses in 1 and 2 will have the form

Component 1   *Component* 2

$$\sigma_{r1} = A_1 \qquad \sigma_{r2} = A_2 - \frac{B_2}{r^2}$$

$$\sigma_{\theta 1} = A_1 \qquad \sigma_{\theta 2} = A_2 + \frac{B_2}{r^2} \tag{10.43}$$

$$\sigma_{z1} = C_1 \qquad \sigma_{z2} = C_2$$

The following boundary conditions exist for our problem:

1. At the interface $r = a$, $\sigma_{r1} = \sigma_{r2}$ for stress continuity.
2. At the free surface $r = b$, $\sigma_{r2} = 0$.
3. The resultant of axial stress σ_z on a section $z =$ constant is zero.
4. Radial displacements in the two components are equal at the interface; that is, at $r = a$, $u_{r1} = u_{r2}$.
5. The radial displacement in component 1 must vanish at the symmetry axis; that is, at $r = 0$, $u_{r1} = 0$.

Using these boundary conditions, it is possible to determine the constants given in Eq. (10.43). The final equations for the matrix sleeve are given below [23, 24]:

$$\sigma_r = A\left(1 - \frac{b^2}{r^2}\right) \qquad \sigma_\theta = A\left(1 + \frac{b^2}{r^2}\right) \qquad \sigma_z = B \tag{10.44}$$

where

$$A = -\left[\frac{E_m(\alpha_m - \alpha_f)\Delta T(a/b)^2}{1 + (a/b)^2(1 - 2v)[(b/a)^2 - 1]E_m/E_f}\right]$$

$$B = \frac{A}{(a/b)^2}\left[2v\left(\frac{a}{b}\right)^2 + \frac{1 + (a/b)^2(1 - 2v) + (a/b)^2(1 - 2v)[(b/a)^2 - 1]E_m/E_f}{1 + [(b/a)^2 - 1]E_m/E_f}\right]$$

$$v_m = v_f = v$$

A plot of σ_r, σ_θ, and σ_z against r/a, where r is the distance in the radial direction and a is the fiber radius, is shown in Fig. 10.9 for the system W/Cu for two fiber volume fractions [25]. Note the change in stress level of σ_z with V_f. This thermoelastic solution can provide information about the magnitude of the elastic stresses involved and whether or not the elastic state will be exceeded. Also, if the matrix deforms plastically in response to these thermal stresses, the above equations can tell where the plastic deformation will initiate. Chawla and Metzger [28] and Chawla [22, 23, 26–28] in a series of studies with metal matrix composites showed that the magnitude of the thermal stresses generated is indeed large enough to deform the soft metallic matrix plastically. Depending on the temperatures involved, the plastic deformation could involve slip, cavitation, grain boundary slinding, and/or migration. They measured the dislocation densities in the copper matrix of tungsten filament/copper single-crystal composites by etch-pitting technique and showed that the dislocation densities were higher near the fiber/matrix interface than away

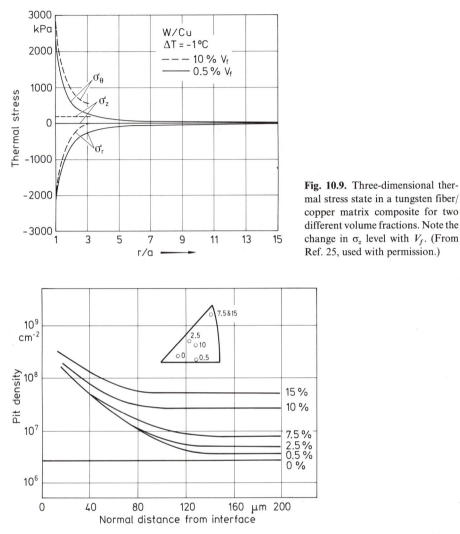

Fig. 10.9. Three-dimensional thermal stress state in a tungsten fiber/copper matrix composite for two different volume fractions. Note the change in σ_z level with V_f. (From Ref. 25, used with permission.)

Fig. 10.10. Variation of dislocation density ($\simeq$ pit density) with distance from the interface. The higher dislocation density in the plateau region with high V_f is due to a higher σ_z with high V_f. (Fig. 10.9). (From Ref. 25, used with permission.)

from the interface, indicating that the plastic deformation, in response to thermal stresses, initiated at the interface. Figure 10.10 shows the variation of dislocation density ($\simeq$ etch pit density) versus distance from the interface. The increase in the dislocation density in the plateau region with V_f (in Fig. 10.10) is due to a higher σ_z with higher V_f value (see Fig. 10.9). Tresca or von Mises yield criteria applied to the stress situation existing in Fig. 10.9 will show that the matrix plastic flow starts at the interface. Further evidence of dislocation generation resulting from thermal stresses has been obtained by Arsenault and coworkers [29, 30] in SiC/Al composites

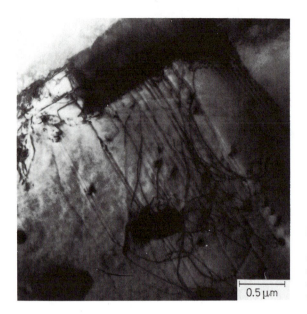

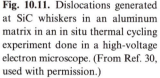

Fig. 10.11. Dislocations generated at SiC whiskers in an aluminum matrix in an in situ thermal cycling experiment done in a high-voltage electron microscope. (From Ref. 30, used with permission.)

by means of transmission electron microscopy. Figure 10.11 shows dislocations generated at SiC whiskers in an aluminum matrix in an in situ thermal cycling experiment done in a high-voltage electron microscope.

10.4 Mechanics of Load Transfer from Matrix to Fiber [31–33]

The matrix holds the fibers together and transmits the applied load to the fibers, the real load-bearing component. Let us focus our attention on a high-modulus fiber embedded in a low-modulus matrix. Figure 10.12a shows the situation prior to the application of an external load. We assume that the fiber and matrix are perfectly bonded and also that the Poisson ratios of the two are the same. Imagine lines running through the fiber/matrix interface in a continuous manner in the unstressed state as shown in Fig. 10.12a. Now let us load this composite axially as shown in Fig. 10.12b. No direct loading of the fibers is permitted. Then the fiber and the matrix experience locally different axial displacements because of the different elastic moduli of the components. Different axial displacements in the

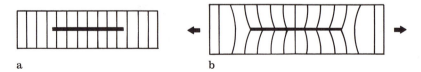

a   b

Fig. 10.12. A high-modulus fiber embedded in a low-modulus matrix: **a** before deformation, **b** after deformation

fiber and the matrix mean shear strains are being produced in the matrix on planes parallel to the fiber axis and in a direction parallel to the fiber axis. Under such circumstances, our imaginary vertical lines of the unstressed state will become distorted as shown in Fig. 10.12b. Transfer of the applied load to the fiber thus occurs by means of these shear strains in the matrix. It is instructive to examine the stress distribution along the fiber/matrix interface. There are two important cases: (a) both the matrix and fiber are elastic and (b) the matrix is plastic and the fiber is elastic. Fibers such as boron, carbon, and ceramic fibers are essentially elastic right up to fracture. Metallic matrices show elastic and plastic deformation before fracture, while polymeric and ceramic matrices can be treated, for all practical purposes, as elastic up to fracture.

10.4.1 Fiber Elastic–Matrix Elastic

Consider a fiber of length l embedded in a matrix subjected to a strain; see Fig. 10.13. We assume that (a) there exists a perfect bonding between the fiber and matrix (i.e., there is no sliding between them) and (b) the Poisson ratios of the fiber and matrix are equal, which implies an absence of transverse stresses when the load is applied along the fiber direction. Let the displacement of a point at a distance x from one extremity of the fiber be u in the presence of a fiber and v in the absence of a fiber. Then we can write for the transfer of load from the matrix to the fiber

$$\frac{dP_f}{dx} = B(u - v) \tag{10.45}$$

where P_f is the load on the fiber and B is a constant that depends on the geometric arrangement of fibers, the matrix type, and the moduli of the fiber and matrix. Differentiating Eq. (10.45), we get

$$\frac{d^2 P_f}{dx^2} = B\left(\frac{du}{dx} - \frac{dv}{dx}\right) \tag{10.46}$$

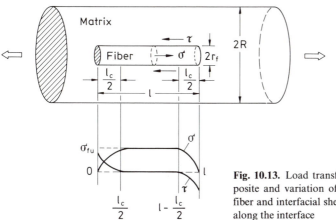

Fig. 10.13. Load transfer in a fiber/matrix composite and variation of tensile stress (σ_f) in the fiber and interfacial shear stress (τ) with distance along the interface

We have

$$\frac{du}{dx} = \text{strain in fiber} = \frac{P_f}{E_f A_f}$$

$$\frac{dv}{dx} = \text{strain in the matrix away from the fiber}$$

$$= \text{imposed strain, } e \ .$$

Thus, Eq. (10.46) can be rewritten

$$\frac{d^2 P_f}{dx^2} = B\left(\frac{P_f}{A_f E_f} - e\right) \tag{10.47}$$

A solution of this differential equation is

$$P_f = E_f A_f e + S \sinh \beta x + T \cosh \beta x \tag{10.48}$$

where

$$\beta = \left(\frac{B}{A_f E_f}\right)^{1/2} \tag{10.49}$$

We use the following boundary condition to evaluate the constants S and T:

$$P_f = 0 \quad \text{at } x = 0 \quad \text{and} \quad x = l$$

Putting in these values and using the half-angle trigonometric formulas, we get the following result:

$$P_f = E_f A_f e \left[1 - \frac{\cosh \beta(l/2 - x)}{\cosh(\beta l/2)}\right] \quad \text{for } 0 < x < l/2 \tag{10.50}$$

or

$$\sigma_f = \frac{P_f}{A_f} = E_f e \left[1 - \frac{\cosh \beta(l/2 - x)}{\cosh(\beta l/2)}\right] \quad \text{for } 0 < x < l/2 \tag{10.51}$$

The maximum possible value of strain in the fiber is the imposed strain e, and thus the maximum stress is eE_f. Therefore, if we have a long enough fiber, the stress in the fiber will increase from the two ends to a maximum value, $\sigma_{fu} = E_f e$. It can be shown readily that the average stress in the fiber is

$$\bar{\sigma}_f = \frac{E_f e}{l} \int_0^l \left[1 - \frac{\cosh \beta(l/2 - x)}{\cosh(\beta l/2)}\right] dx = E_f e \left[1 - \frac{\tanh(\beta l/2)}{\beta l/2}\right] \tag{10.52}$$

We can obtain the variation of shear stress τ along the fiber/matrix interface by considering the equilibrium of forces acting over an element of fiber (radius r_f). Thus, we can write from Fig. 10.13

$$\frac{dP_f}{dx} dx = 2\pi r_f \, dx \, \tau \tag{10.53}$$

Now P_f, the tensile load on the fiber, is equal to $\pi r_f^2 \sigma_f$. Substituting this in Eq. (10.53), we get

$$\tau = \frac{1}{2\pi r_f} \frac{dP_f}{dx} = \frac{r_f}{2} \frac{d\sigma_f}{dx} \tag{10.54}$$

From Eqs. (10.51) and (10.54), we obtain

$$\tau = \frac{E_f r_f e \beta}{2} \frac{\sinh \beta(l/2 - x)}{\cosh(\beta l/2)} \tag{10.55}$$

Figure 10.13 shows the variation of τ and σ_f with distance x. The maximum shear stress, in Eq. (10.55), will be the smaller of the following two shear stresses: namely, (a) the shear yield stress of the matrix or (b) the shear strength of the fiber/matrix interface. Whichever of these two shear stresses is attained first will control the load transfer phenomenon and should be used in Eq. (10.55).

Now we can determine the constant B. The value of B depends on the fiber packing geometry. Consider Fig. 10.13 again and let the fiber length l be much greater than the fiber radius r_f. Let $2R$ be the average fiber spacing (center to center). Let us also denote the shear stress in the fiber direction at a distance r from the axis by $\tau(r)$. Then, at the fiber surface ($r = r_f$), we can write

$$\frac{dP_f}{dx} = -2\pi r_f \tau(r_f) = B(u - v)$$

Thus,

$$B = -\frac{2\pi r_f \tau(r_f)}{u - v} \tag{10.56}$$

Let w be the real displacement in the matrix. Then at the fiber/matrix interface, no sliding being permitted, $w = u$. At a distance R from the center of a fiber, the matrix displacement is unaffected by the fiber presence and $w = v$. Considering now the equilibrium of forces acting on the matrix volume between r_f and R, we can write

$$2\pi r \tau(r) = \text{const} = 2\pi r_f \tau(r_f)$$

or

$$\tau(r) = \frac{\tau(r_f) r_f}{r} \tag{10.57}$$

The shear strain γ in the matrix is given by $\tau(r) = G_m \gamma$, where G_m is the matrix shear modulus. Then

$$\gamma = \frac{dw}{dr} = \frac{\tau(r)}{G_m} = \frac{\tau(r_f) r_f}{G_m r} \tag{10.58}$$

Integrating from r_f to R, we obtain

$$\int_{r_f}^{R} dw = \Delta w = \frac{\tau(r_f) r_f}{G_m} \int_{r_f}^{R} \frac{1}{r} dr = \frac{\tau(r_f) r_f}{G_m} \ln\left(\frac{R}{r_f}\right) \tag{10.59}$$

But, by definition,

$$\Delta w = v - u = -(u - v) \tag{10.60}$$

From Eqs. (10.59) and (10.60), we get

$$\frac{\tau(r_f)r_f}{u-v} = -\frac{G_m}{\ln(R/r_f)} \tag{10.61}$$

From Eqs. (10.56) and (10.61), one obtains

$$B = \frac{2\pi G_m}{\ln(R/r_f)} \tag{10.62}$$

and from Eq. (10.49), one can obtain an expression for the load transfer parameter β:

$$\beta = \left(\frac{B}{E_f A_f}\right)^{1/2} = \left[\frac{2\pi G_m}{E_f A_f \ln(R/r_f)}\right]^{1/2} \tag{10.63}$$

The value of R/r_f is a function of fiber packing. For a square array of fibers $\ln(R/r_f) = \frac{1}{2}\ln(\pi/V_f)$, while for a hexagonal packing $\ln(R/r_f) = \frac{1}{2}\ln(2\pi/\sqrt{3}V_f)$. We can define $\ln(R/r_f) = \frac{1}{2}\ln(\phi_{max}/V_f)$, where ϕ_{max} is the maximum packing factor. Substituting this in Eq. (10.63), we get

$$\beta = \left[\frac{4\pi G_m}{E_f A_f \ln(\phi_{max}/V_f)}\right]^{1/2}$$

Note that the greater the value of the ratio G_m/E_f, the greater is the value of β and the more rapid is the stress increase in the fiber from either end.

More rigorous analyses give results similar to the one above and differ only in the value of β. In all analyses, β is proportional to $(G_m/E_f)^{1/2}$, and differences occur only in the term involving fiber volume fraction, $\ln(R/r_f)$, in the above equation.

10.4.2 Fiber Elastic–Matrix Plastic

It should be clear from the discussion above that to load high-strength fibers in a ductile matrix to their maximum strength, the matrix shear strength must be large. A metallic matrix will flow plastically in response to the high shear stresses developed. Of course, if the fiber/matrix interface is weaker, it will fail first. Assuming that the plastically deforming matrix does not work harden, the shear stress at the fiber surface, $\tau(r_f)$, will have an upper limit of τ_y, the matrix shear yield strength. In PMCs and CMCs, frictional slip at the interface is more likely than plastic flow of the matrix. In the case of PMCs and CMCs, therefore, the limiting shear stress will be the interface strength in shear, τ_i. The term τ_i should replace τ_y in what follows for PMCs and CMCs. If the polymer shrinkage during curing results in a radial pressure p on the fibers, then τ_y should be replaced by μp as $\tau_i = \mu p$, where μ is the coefficient of sliding friction between the fiber and matrix [31]. The equilibrium of forces then gives

$$\sigma_f \frac{\pi d^2}{4} = \tau_y \pi d \frac{l}{2}$$

or

$$\frac{l}{d} = \frac{\sigma_f}{2\tau_y}$$

We consider $l/2$ and not l because the fiber is being loaded from both ends. Given a sufficiently long fiber, it should be possible to load it to its breaking stress σ_{fu} by means of load transfer through the matrix flowing plastically around it. Let $(l/d)_c$ be the maximum fiber length to diameter ratio necessary to accomplish this. We call this ratio (l/d) the aspect ratio of a fiber and $(l/d)_c$ is the critical aspect ratio necessary to attain the breaking stress of the fiber, σ_{fu}. Then we can write

$$\left(\frac{l}{d}\right)_c = \frac{\sigma_{fu}}{2\tau_y} \tag{10.64}$$

For a given fiber diameter d, we can think of a critical fiber length l_c. Thus,

$$\frac{l_c}{d} = \frac{\sigma_{fu}}{2\tau_y}. \tag{10.65}$$

Over a length l_c, the load in the fiber builds up from both ends. Strain builds in a likewise manner. Beyond l_c (i.e., in the middle portion of the fiber), the local displacements in the matrix and fiber are the same, and the fiber carries the major load while the matrix carries only a minor portion of the applied load. Equation (10.65) tells us that the fiber length l must be equal or greater than l_c for the fiber to be loaded to its maximum stress, σ_{fu}. For $l < l_c$, the matrix will flow plastically around the fiber and will load it to a stress in its central portion given by

$$\sigma_f = 2\tau\frac{l}{d} < \sigma_{fu} \tag{10.66}$$

This is shown in Fig. 10.14. An examination of this figure shows that even for $l/d > (l/d)_c$ the average stress in the fiber will be less than the maximum stress to which it is loaded in its central region. In fact, one can write for the average fiber stress

$$\bar{\sigma}_f = \frac{1}{l}\int_0^l \sigma_f \, dx = \frac{1}{l}[\sigma_f(l - l_c) + \beta\sigma_f l_c] = \frac{1}{l}[\sigma_f l - l_c(\sigma_f - \beta\sigma_f)]$$

or

$$\bar{\sigma}_f = \sigma_f\left(1 - \frac{1 - \beta}{l/l_c}\right) \tag{10.67}$$

where $\beta\sigma_f$ is the average stress in the fiber over a portion $l_c/2$ of its length at both

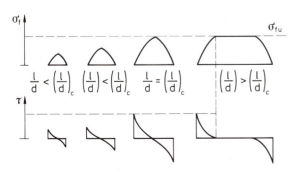

Fig. 10.14. Schematic of variation of tensile stress in a fiber (σ_f) and interface shear stress (τ) with different fiber aspect ratios (l/d)

the ends. One can thus regard β as a load transfer function. Its value will be precisely 0.5 for an ideally plastic material; that is, the increase in stress in the fiber over the portion $l_c/2$ will be linear. The composite stress can then be written per the rule of mixtures:

$$\sigma_c = \sigma_f V_f \left(1 - \frac{1-\beta}{l/l_c}\right) + \sigma_m'(1 - V_f) \tag{10.68}$$

If $\beta = 0.5$,

$$\sigma_c = \sigma_f V_f \left(1 - \frac{l_c}{2l}\right) + \sigma_m'(1 - V_f)$$

where σ_m' is the in situ matrix stress at the strain under consideration. Suppose now that in a whisker reinforced metal the whiskers have an $l/l_c = 10$; then it can easily be shown that the strength of such a composite containing discontinuous but aligned fibers will be 95% of that of a composite containing continuous fibers. Thus, as long as the fibers are reasonably long compared to the load transfer length, there is not much loss of a strength owing to their discontinuous nature. The stress concentration effect at the ends of the discontinuous fibers has been neglected in this simple analysis.

References

1   W. Voigt, *Lehrbuch der Kristallphysik*, Teubner, Leipzig, 1910.
2   A. Reuss, *Z. Angew. Math. Mech.*, **9**, 49 (1929).
3   C.C. Chamis and G.P. Sendeckyj, *J. Composite Mater.*, **2**, 332 (1968).
4   J.F. Nye, *Physical Properties of Crystals*, Oxford University Press, London, 1969, p. 131.
5   Z. Hashin and B.W. Rosen, *J. Appl. Mech.*, **31**, 233 (1964), ASME.
6   B.W. Rosen, *Composites*, **4**, 16 (Jan. 1973).
7   C.C. Chamis, NASA Tech. Memo. 83320, 1983 [presented at the 38th Annual Conference of the Society of Plastics Industry (SPI), Houston, TX, Feb. 1983].
8   R. Hill, *J. Mech. Phys. Solids*, **12**, 199 (1964).
9   R. Hill, *J. Mech. Phys. Solids*, **13**, 189 (1965).
10   J.C. Halpin and S.W. Tsai, "Environmental Factors Estimation in Composite Materials Design," AFML TR 67-423 (1967).
11   J.C. Halpin and J.L. Kardos, *Polym. Eng. Sci.*, **16**, 344 (1976).
12   J.L. Kardos, *CRC Crit. Rev. Solid State Sci.*, **3**, 419 (1971).
13   J.M. Whitney, *J. Structural Div. Am. Soc. Civil Eng.*, 113 (Jan. 1973).
14   L.E. Nielsen, *Mechanical Properties of Polymers and Composites*, vol. 2, Marcel Dekker, New York, 1974.
15   A. Kelly, in *Chemical and Mechanical Behavior of Inorganic Materials*, Wiley-Interscience, New York, 1970, p. 523.
16   A.E.H. Love, *A Treatise on the Mathematical Theory of Elasticity*, 4th ed., Dover, New York, p. 144.
17   A. Kelly and H. Lilholt, *Philos. Mag.*, **20**, 175 (1971).
18   P.S. Turner, *J. Res. Natl. Bur. Stand.*, **37**, 239 (1946).
19   E.H. Kerner, *Proc. Phys. Soc. London*, **B69**, 808 (1956).
20   R.A. Schapery, *J. Composite Mater.*, **2**, 311, 1969.
21   E. Behrens, *J. Composite Mater.*, **2**, 2 (1968).
22   K.K. Chawla, *Philos. Mag.*, **28**, 401 (1973).

23   K.K. Chawla, *Metallography*, **6**, 155 (1973).
24   H. Poritsky, *Physics*, **5**, 406 (1934).
25   K.K. Chawla and M. Metzger, *J. Mater. Sci.*, **7**, 34 (1972).
26   K.K. Chawla, in *Proceedings of the International Conference on Composite Materials/1975*, TMS-AIME, New York, 1976, p. 535.
27   K.K. Chawla, *J. Mater. Sci.*, **11**, 1567 (1976).
28   K.K. Chawla, in *Grain Boundaries in Engineering Materials*, Claitor's Publishing Division, Baton Rouge, LA, 1974, p. 435.
29   R.J. Arsenault and R.M. Fisher, *Scripta Met.*, **17**, 67 (1983).
30   M. Vogelsang, R.J. Arsenault, and R.M. Fisher, *Met. Trans. A*, **7A**, 379 (1986).
31   A. Kelly, *Strong Solids*, 2nd ed., Clarendon Press, Oxford, 1973, p. 157.
32   D.M. Schuster and E. Scala, *Trans. Met. Soc.-AIME*, **230**, 1635 (1964).
33   N.F. Dow, General Electric Report No. R63-SD-61, 1963.

Suggested Reading

B.D. Agarwal and L.J. Broutman, *Analysis and Performance of Fiber Composites*, John Wiley & Sons, New York, 1980.

D. Hull, *An Introduction to Composite Materials*, Cambridge University Press, Cambridge, U.K., 1981.

A. Kelly, *Strong Solids*, 2nd ed., Clarendon Press, Oxford, 1973.

M.R. Piggott, *Load-Bearing Fibre Composites*, Pergamon Press, Oxford, 1980.

V.K. Tewary, *Mechanics of Fibre Composites*, Halsted Press, New York, 1978.

11. Macromechanics of Composites

Laminated fibrous composites are made by bonding together two or more laminae. The individual unidirectional laminae or plies are oriented in such a manner that the resulting structural component has the desired mechanical and/or physical characteristics in different directions. Thus, one exploits the inherent anisotropy of fibrous composites to design a composite material having the appropriate properties.

In Chap. 10 we treated the micromechanics of fibrous composites, that is, how to obtain the composite properties when the properties of the matrix and the fiber as well as their geometric arrangements are known. While micromechanics is very useful in analyzing the composite behavior, we use the information obtained from a micromechanical analysis of a thin unidirectional lamina (or in case of a lack of such analytical information, we must determine experimentally the properties of a lamina) as input for a macromechanical analysis of a laminated composite. Figure 11.1 shows this concept schematically [1]. Once we have determined, analytically or otherwise, the characteristics of a fibrous lamina, we ignore its detailed microstructural nature and simply treat it as a homogeneous, orthotropic sheet. A laminated composite is made by stacking a number of such orthotropic sheets at specific orientations to get the composite materials having desired characteristics. We then use the existing theory of laminated plates or shells to analyze such laminated composites.

To appreciate the significance of such a macromechanical analysis, we first review the basic ideas of the elastic constants of a bulk isotropic material and a lamina, a lamina as an orthotropic sheet, and finally the use of classical laminated

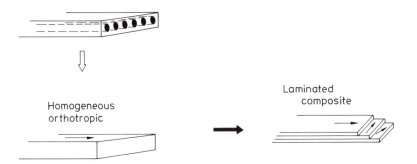

Homogeneous
orthotropic

Laminated
composite

Fig. 11.1. Macromechanical analysis of laminate composites. A unidirectional ply is treated as a homogeneous, orthotropic material. Many such plies are stacked in an appropriate order (following laminated plate or shell theory) to make the composite. (Reprinted from Ref. 1, by Courtesy of Marcel Dekker, Inc.)

plate theory to analyze macromechanically the laminated composites. The reader is referred to some standard texts on elasticity [2–4] for a detailed review on the subject.

11.1 Elastic Constants of an Isotropic Material

Stress is defined as the force per unit area of a body. We can represent the stresses acting at a point in a solid by the stresses acting on the surfaces of an elemental cube at that point. There are nine stress components acting on the front faces of the elemental cube. The component σ_{ij} represents the force per unit area in the i direction on a face whose normal is the j direction. Rotational equilibrium requires that $\sigma_{ij} = \sigma_{ji}$. Thus, we are left with six stress components; $i = j$ are the normal stresses while $i \neq j$ are the shear stresses.

The displacement of a point in a deformed body with respect to its original position in the undeformed state can be represented by a vector $\mathbf{u}$ with components u_1, u_2, and u_3; these components are the projections of $\mathbf{u}$ on the x_1, x_2, and x_3 axes. Strain is defined as the ratio of change in length to original length. We can define the strain components in terms of the first derivatives of the displacement components as follows:

$$\varepsilon_{ij} = \frac{1}{2}\left(\frac{\partial u_i}{\partial x_j} + \frac{\partial u_j}{\partial x_i}\right).$$

Here $i \neq j$ represents shear strains and $i = j$ gives normal strains. It should be noted that ε_{ij}, for $i \neq j$, gives one-half the engineering shear strain, γ_{ij}; that is

$$\gamma_{ij} = 2\varepsilon_{ij} = \frac{\partial u_i}{\partial x_j} + \frac{\partial u_j}{\partial x_i}.$$

The relationship between stress and strain in linear elasticity is described by Hooke's law. Hooke's law states that, for small strains, the stress is proportional to the strain. For the simple case of a stress applied unidirectionally to an isotropic solid, we can write Hooke's law as

$$\sigma = E\varepsilon \tag{11.1}$$

where σ is the unidirectional stress, ε is the strain in the applied stress direction, and E is Young's modulus. We have omitted the indices in this simple unidirectional case.

In its most generalized form, Hooke's law can be written

$$\sigma_{ij} = C_{ijkl}\varepsilon_{kl} \tag{11.2}$$

where C_{ijkl} are the elastic constants or stiffnesses. Equation (11.2) is in tensor form. Written out in the expanded form, it will have 81 elastic constants. It is general practice to use a contracted matrix notation for writing stresses, strains, and elastic constants.

We use C_{mn} for C_{ijkl}, σ_m for σ_{ij}, and ε_n for ε_{kl} as per the following procedure:

ij or kl	11	22	33	23	31	12
m or n	1	2	3	4	5	6

Then Eq. (11.2) can be rewritten

$$\sigma_m = C_{mn}\varepsilon_n . \tag{11.3}$$

It can be shown that $C_{mn} = C_{nm}$.
Conversely, we can write

$$\varepsilon_m = S_{mn}\sigma_n \tag{11.4}$$

where S_{mn}, the compliance matrix, is the inverse of the stiffness matrix C_{mn}.
In the expanded form, we have

$$
\begin{bmatrix} \sigma_1 \\ \sigma_2 \\ \sigma_3 \\ \sigma_4 \\ \sigma_5 \\ \sigma_6 \end{bmatrix}
=
\begin{bmatrix}
C_{11} & C_{12} & C_{13} & C_{14} & C_{15} & C_{16} \\
 & C_{22} & C_{23} & C_{24} & C_{25} & C_{26} \\
 & & C_{33} & C_{34} & C_{35} & C_{36} \\
 & & & C_{44} & C_{45} & C_{46} \\
 & & & & C_{55} & C_{56} \\
 & & & & & C_{66}
\end{bmatrix}
\begin{bmatrix} \varepsilon_1 \\ \varepsilon_2 \\ \varepsilon_3 \\ \varepsilon_4 \\ \varepsilon_5 \\ \varepsilon_6 \end{bmatrix}
\tag{11.5}
$$

Note that σ_4, σ_5, and σ_6 now represent the shear stresses while ε_4, ε_5, and ε_6 represent the engineering shear strains. The dashed line along the diagonal indicates that the matrix is symmetrical. Equation (11.5) thus gives 21 independent elastic constants in the most general case.

For most materials, the number of independent elastic constants is further reduced because of the various symmetry elements present. For example, only three elastic constants are independent for cubic systems. For isotropic materials where elastic properties are independent of direction, only two constants are independent. For isotropic materials, Eq. (11.5) reduces to

$$
\begin{bmatrix} \sigma_1 \\ \sigma_2 \\ \sigma_3 \\ \sigma_4 \\ \sigma_5 \\ \sigma_6 \end{bmatrix}
=
\begin{bmatrix}
C_{11} & C_{12} & C_{12} & 0 & 0 & 0 \\
 & C_{11} & C_{12} & 0 & 0 & 0 \\
 & & C_{11} & 0 & 0 & 0 \\
 & & & \dfrac{C_{11} - C_{12}}{2} & 0 & 0 \\
 & & & & \dfrac{C_{11} - C_{12}}{2} & 0 \\
 & & & & & \dfrac{C_{11} - C_{12}}{2}
\end{bmatrix}
\begin{bmatrix} \varepsilon_1 \\ \varepsilon_2 \\ \varepsilon_3 \\ \varepsilon_4 \\ \varepsilon_5 \\ \varepsilon_6 \end{bmatrix}
\tag{11.6}
$$

In terms of the compliance matrix, we can write for isotropic materials

$$
\begin{bmatrix} \varepsilon_1 \\ \varepsilon_2 \\ \varepsilon_3 \\ \varepsilon_4 \\ \varepsilon_5 \\ \varepsilon_6 \end{bmatrix} = \begin{bmatrix} S_{11} & S_{12} & S_{12} & 0 & 0 & 0 \\ & S_{11} & S_{12} & 0 & 0 & 0 \\ & & S_{11} & 0 & 0 & 0 \\ & & & 2(S_{11} - S_{12}) & 0 & 0 \\ & & & & 2(S_{11} - S_{12}) & 0 \\ & & & & & 2(S_{11} - S_{12}) \end{bmatrix} \begin{bmatrix} \sigma_1 \\ \sigma_2 \\ \sigma_3 \\ \sigma_4 \\ \sigma_5 \\ \sigma_6 \end{bmatrix}
\tag{11.7}
$$

Only C_{11} and C_{12} (or S_{11} and S_{12}) are the independent constants. Engineers frequently use elastic constants such as Young's modulus E, Poisson ratio v, shear modulus G, and bulk modulus K. Only two of these are independent because E, G, v, and K are interrelated:

$$
E = 2G(1 + v) \quad \text{and} \quad K = \frac{E}{3(1 - 2v)} \; .
$$

The relationships between these engineering constants and compliances are as follows:

$$
E = \frac{1}{S_{11}} \qquad v = -\frac{S_{12}}{S_{11}} \qquad G = (\tfrac{1}{2})(S_{11} - S_{12})
$$

and the compliances are related to the stiffnesses as follows:

$$
S_{11} = \frac{C_{11} + C_{12}}{(C_{11} - C_{12})(C_{11} + 2C_{12})}
$$

$$
S_{12} = -\frac{C_{12}}{(C_{11} - C_{12})(C_{11} + 2C_{12})}
$$

11.2 Elastic Constants of a Lamina

We can make a laminate composite by stacking up a sufficiently large number of laminae. A lamina, the unit building block of a composite, can be considered to represent a state of generalized plane stress. This implies that the through thickness stress components are zero. Thus $\sigma_3 = \sigma_4 = \sigma_5 = 0$. Then Eqs. (11.6) and (11.7) are reduced, for an isotropic lamina, to

$$
\begin{bmatrix} \sigma_1 \\ \sigma_2 \\ \sigma_6 \end{bmatrix} = \begin{bmatrix} C_{11} & C_{12} & 0 \\ C_{12} & C_{11} & 0 \\ 0 & 0 & \dfrac{C_{11} - C_{12}}{2} \end{bmatrix} \begin{bmatrix} \varepsilon_1 \\ \varepsilon_2 \\ \varepsilon_6 \end{bmatrix}
\tag{11.8}
$$

$$
\begin{bmatrix} \varepsilon_1 \\ \varepsilon_2 \\ \varepsilon_6 \end{bmatrix} = \begin{bmatrix} S_{11} & S_{12} & 0 \\ S_{12} & S_{11} & 0 \\ 0 & 0 & 2(S_{11} - S_{12}) \end{bmatrix} \begin{bmatrix} \sigma_1 \\ \sigma_2 \\ \sigma_6 \end{bmatrix}
\tag{11.9}
$$

Equations (11.8) and (11.9) describe the stress–strain relationships for an isotropic lamina, for example, an aluminum sheet. A fiber reinforced lamina, however, is not an isotropic material. It is an orthotropic material; that is, it has three mutually perpendicular axes of symmetry. Relationships become slightly more complicated when we have orthotropy rather than isotropy.

A fiber reinforced lamina or ply is a thin sheet (~ 0.1 mm) containing oriented fibers. Generally, the fibers are oriented unidirectionally as in a prepreg but fibers in the form of a woven roving may also be used. Several such thin laminae are stacked in a specific order of fiber orientation, cured, and bonded into a laminated composite.

Since the behavior of a laminated composite depends on the characteristics of individual laminae, and with due regard to their directionality, we discuss now the elastic behavior of an orthotropic lamina.

For the case of an orthotropic material with the coordinate axes parallel to the symmetry axes of the material, the array of elastic constants is given by

$$[S_{ij}] = \begin{bmatrix} S_{11} & S_{12} & S_{13} & 0 & 0 & 0 \\ & S_{22} & S_{23} & 0 & 0 & 0 \\ & & S_{33} & 0 & 0 & 0 \\ & & & S_{44} & 0 & 0 \\ & & & & S_{55} & 0 \\ & & & & & S_{66} \end{bmatrix} \tag{11.10}$$

A similar expression can be written for C_{ij}. Taking into account the fact that a lamina is a thin orthotropic material, that is, through thickness components are zero, we can write the stiffness matrix for an orthotropic lamina by eliminating the terms involving the z axis:

$$[S_{ij}] = \begin{bmatrix} S_{11} & S_{12} & 0 \\ & S_{22} & 0 \\ & & S_{66} \end{bmatrix} \tag{11.11}$$

We can rewrite in full form Hooke's law for a thin orthotropic lamina, with natural and geometric axes coinciding, as follows:

$$\begin{bmatrix} \varepsilon_1 \\ \varepsilon_2 \\ \varepsilon_6 \end{bmatrix} = \begin{bmatrix} S_{11} & S_{12} & 0 \\ S_{12} & S_{22} & 0 \\ 0 & 0 & S_{66} \end{bmatrix} \begin{bmatrix} \sigma_1 \\ \sigma_2 \\ \sigma_6 \end{bmatrix} \tag{11.12}$$

Conversely,

$$\begin{bmatrix} \sigma_1 \\ \sigma_2 \\ \sigma_6 \end{bmatrix} = \begin{bmatrix} Q_{11} & Q_{12} & 0 \\ Q_{12} & Q_{22} & 0 \\ 0 & 0 & Q_{66} \end{bmatrix} \begin{bmatrix} \varepsilon_1 \\ \varepsilon_2 \\ \varepsilon_6 \end{bmatrix} \tag{11.13}$$

It is customary to use the symbol Q_{ij} rather than C_{ij} for thin material. The Q_{ij} are called reduced stiffnesses. The relationships between Q_{ij} and S_{ij} can easily be shown to be

$$Q_{11} = \frac{S_{22}}{S_{11}S_{22} - S_{12}^2}$$

$$Q_{12} = -\frac{S_{12}}{S_{11}S_{22} - S_{12}^2}$$

$$Q_{22} = \frac{S_{11}}{S_{11}S_{22} - S_{12}^2}$$

(11.14)

$$Q_{66} = \frac{1}{S_{66}}$$

Also,

$$Q_{ij} = C_{ij} - \frac{C_{i3}C_{j3}}{C_{33}} \qquad (i,j = 1,2,6)$$

Note that three-dimensional orthotropy requires nine independent elastic constants [Eq. (11.10)], while bidimensional orthotropy requires only four [Eq. (11.11)]. For an isotropic material (two or three dimensional) one just needs two indepenent elastic constants [Eqs. (11.6) and (11.8)].

It is worth emphasizing that Eqs. (11.12) and (11.13), showing terms with indices 16 and 26 to be zero, represent a special case of orthotropy when the principal material axes of symmetry (the fiber direction [1] and the direction transverse to it [2]) coincide with the principal loading direction. If this is not so, that is, if the material symmetry axes and the geometric axes do not coincide, which is a more general case of orthotropy, then we shall have a fully populated elastic constant matrix and the stress–strain relationships become

$$\begin{bmatrix} \sigma_x \\ \sigma_y \\ \sigma_s \end{bmatrix} = \begin{bmatrix} \bar{Q}_{11} & \bar{Q}_{12} & \bar{Q}_{16} \\ \bar{Q}_{12} & \bar{Q}_{22} & \bar{Q}_{26} \\ \bar{Q}_{16} & \bar{Q}_{26} & \bar{Q}_{66} \end{bmatrix} \begin{bmatrix} \varepsilon_x \\ \varepsilon_y \\ \varepsilon_s \end{bmatrix}$$

(11.15)

where the $\bar{Q}_{ij}$ matrix is called the transformed reduced stiffness matrix because it is obtained by transforming Q_{ij} (specially orthotropic) to $\bar{Q}_{ij}$ (generally orthotropic). We can perform the transformation of axes as shown below and obtain $\bar{Q}_{ij}$ from Q_{ij}.

Figure 11.2 shows the situation for a unidirectional composite lamina where the two sets of axes do not coincide. The properties in the 1-2 system are known, and we wish to determine them in the x-y system or vice versa. Both stress and strain are second-rank tensors. A second-rank tensor, T_{ij}, transforms as

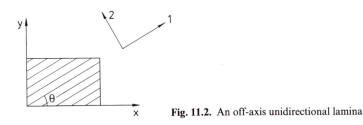

Fig. 11.2. An off-axis unidirectional lamina

Table 11.1. Direction cosines

Direction	x	y	
1	$a_{11} = m$	$a_{12} = n$	$m = \cos\theta$
2	$a_{21} = -n$	$a_{22} = m$	$n = \sin\theta$

$$T_{ij} = a_{ik}a_{jl}T_{kl}$$

where a_{ik} and a_{jl} are the direction cosines. Table 11.1 gives the direction cosines for the axes transformation shown in Fig. 11.2. Angle θ is positive when the x-y axes are rotated counterclockwise with respect to the 1-2 axes. This transformation of axes is carried out easily in the matrix form (see Appendix A). For stresses we can write

$$\begin{bmatrix} \sigma_1 \\ \sigma_2 \\ \sigma_6 \end{bmatrix} = [T]_\sigma \begin{bmatrix} \sigma_x \\ \sigma_y \\ \sigma_s \end{bmatrix} \tag{11.16}$$

while for strains,

$$\begin{bmatrix} \varepsilon_1 \\ \varepsilon_2 \\ \varepsilon_6 \end{bmatrix} = [T]_\varepsilon \begin{bmatrix} \varepsilon_x \\ \varepsilon_y \\ \varepsilon_s \end{bmatrix} \tag{11.17}$$

where $[T]_\sigma$ and $[T]_\varepsilon$ are the transformation matrices for stress and strain transformations, respectively, and are given by

$$[T]_\sigma = \begin{bmatrix} m^2 & n^2 & 2mn \\ n^2 & m^2 & -2mn \\ -mn & mn & m^2 - n^2 \end{bmatrix} \tag{11.18}$$

$$[T]_\varepsilon = \begin{bmatrix} m^2 & n^2 & mn \\ n^2 & m^2 & -mn \\ -2mn & 2mn & m^2 - n^2 \end{bmatrix} \tag{11.19}$$

where $m = \cos\theta$ and $n = \sin\theta$. This method of using different transformation matrices for stress and strain transformations avoids the need of putting the factor $\frac{1}{2}$ before the engineering shear strains to convert them to tensorial strain components suitable for transformation. Multiplying both sides of Eq. (11.16) by $[T]_\sigma^{-1}$ and remembering that $[T]_\sigma[T]_\sigma^{-1} = [I]$, the identity matrix, we get

$$\begin{bmatrix} \sigma_x \\ \sigma_y \\ \sigma_s \end{bmatrix} = [T]_\sigma^{-1} \begin{bmatrix} \sigma_1 \\ \sigma_2 \\ \sigma_6 \end{bmatrix} \tag{11.20}$$

$[T]_\sigma^{-1}$ can be obtained from $[T]_\sigma$ by simply substituting $-\theta$ for θ. Appendix A gives the procedure for obtaining the inverse of a given matrix. In this particular case, substituting $-\theta$ for θ in Eq. (11.18) results in

$$[T]_\sigma^{-1} = \begin{bmatrix} m^2 & n^2 & -2mn \\ n^2 & m^2 & 2mn \\ mn & -mn & m^2 - n^2 \end{bmatrix} \tag{11.21}$$

where $m = \cos\theta$ and $n = \sin\theta$. Substituting Eq. (11.13). in Eq. (11.20), we obtain

$$\begin{bmatrix} \sigma_x \\ \sigma_y \\ \sigma_s \end{bmatrix} = [T]_\sigma^{-1}[Q] \begin{bmatrix} \varepsilon_1 \\ \varepsilon_2 \\ \varepsilon_6 \end{bmatrix} \tag{11.22}$$

If we now substitute Eq. (11.17) in Eq. (11.22), we arrive at

$$\begin{bmatrix} \sigma_x \\ \sigma_y \\ \sigma_s \end{bmatrix} = [T]_\sigma^{-1}[Q][T]_\varepsilon \begin{bmatrix} \varepsilon_x \\ \varepsilon_y \\ \varepsilon_s \end{bmatrix} = [\bar{Q}] \begin{bmatrix} \varepsilon_x \\ \varepsilon_y \\ \varepsilon_s \end{bmatrix} \tag{11.23}$$

where

$$[\bar{Q}] = [T]_\sigma^{-1}[Q][T]_\varepsilon . \tag{11.24}$$

$[\bar{Q}]$ is the stiffness matrix for a generally orthotropic lamina whose components in expanded form are written as follows ($m = \cos\theta, n = \sin\theta$):

$$\begin{aligned}
\bar{Q}_{11} &= Q_{11}m^4 + 2(Q_{12} + 2Q_{66})m^2n^2 + Q_{22}n^4 \\
\bar{Q}_{12} &= (Q_{11} + Q_{22} - 4Q_{66})m^2n^2 + Q_{12}(m^4 + n^4) \\
\bar{Q}_{22} &= Q_{11}n^4 + 2(Q_{12} + 2Q_{66})m^2n^2 + Q_{22}m^4 \\
\bar{Q}_{16} &= (Q_{11} - Q_{12} - 2Q_{66})m^3n + (Q_{12} - Q_{22} + 2Q_{66})mn^3 \\
\bar{Q}_{26} &= (Q_{11} - Q_{12} - 2Q_{66})mn^3 + (Q_{12} - Q_{22} + 2Q_{66})m^3n \\
\bar{Q}_{66} &= (Q_{11} + Q_{22} - 2Q_{12} - 2Q_{66})m^2n^2 + Q_{66}(m^4 + n^4)
\end{aligned} \tag{11.25}$$

Note that although $\bar{Q}_{ij}$ is a completely filled matrix, only four of its components are independent: $\bar{Q}_{16}$ and $\bar{Q}_{26}$ are linear combinations of the other four.

A corresponding stress–strain relationship in terms of compliances of a generally orthotropic lamina can be obtained:

$$\begin{bmatrix} \varepsilon_x \\ \varepsilon_y \\ \varepsilon_s \end{bmatrix} = \begin{bmatrix} S_{11} & S_{12} & S_{16} \\ S_{12} & S_{22} & S_{26} \\ S_{16} & S_{26} & S_{66} \end{bmatrix} \begin{bmatrix} \sigma_x \\ \sigma_y \\ \sigma_s \end{bmatrix} \tag{11.26}$$

In a generally orthotropic lamina wherein we have nonzero 16 and 26 terms, a unidirectional normal stress σ_x has both normal as well as shear strains as responses and vice versa: that is, there is a coupling between the normal and shear effects. In the case of a specially orthotropic lamina where the 16 and 26 terms are zero, we have normal stresses producing normal strains and shear stresses producing shear strains and vice versa. In this there is no coupling between the normal and shear components. We shall present more about such coupling effects in composites later.

11.3 Relationships Between Engineering Constants and Reduced Stiffnesses and Compliances

Consider the thin lamina shown in Fig. 11.3 with the natural axes coinciding with the geometric axes. The conventional engineering constants in this case are Young's moduli in direction 1 (E_1) and direction 2 (E_2), the principal shear modulus G_6, and the principal Poisson ratio ν_1. The Poisson ratio ν_1, when the lamina is strained in direction 1, is equal to $-\varepsilon_2/\varepsilon_1$. The Poisson ratio ν_2, when the lamina is strained in direction 2, equals $-\varepsilon_1/\varepsilon_2$.

We wish to relate these five conventional engineering constants to the four independent elastic constants, the reduced stiffnesses Q_{ij}. Let us consider that σ_1 is the only nonzero component in Eq (11.13). Then we can write

$$\sigma_1 = Q_{11}\varepsilon_1 + Q_{12}\varepsilon_2$$

$$\sigma_2 = Q_{12}\varepsilon_1 + Q_{22}\varepsilon_2 = 0$$

Solving for ε_1 and ε_2, we get

$$\varepsilon_1 = \frac{Q_{22}}{Q_{11}Q_{22} - Q_{12}^2}\sigma_1$$

and

$$\varepsilon_2 = -\frac{Q_{12}}{Q_{11}Q_{22} - Q_{12}^2}\sigma_1$$

By definition we have, $E_1 = \sigma_1/\varepsilon_1$. Thus,

$$E_1 = \frac{Q_{11}Q_{22} - Q_{12}^2}{Q_{22}} \tag{11.27}$$

and

$$\nu_1 = -\frac{\varepsilon_2}{\varepsilon_1} = \frac{Q_{12}}{Q_{22}} \tag{11.28}$$

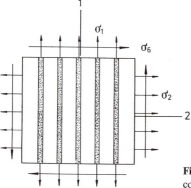

Fig. 11.3. A thin lamina with natural (or material) axes coinciding with the geometric axes

If we repeat the above procedure with σ_2 as the only nonzero stress component in Eq. (11.13), we obtain

$$E_2 = \frac{\sigma_2}{\varepsilon_2} = \frac{Q_{11}Q_{22} - Q_{12}^2}{Q_{11}} \tag{11.29}$$

and

$$\nu_2 = -\frac{\varepsilon_1}{\varepsilon_2} = \frac{Q_{12}}{Q_{11}} \tag{11.30}$$

If we consider that σ_6 is the only nonzero component, we can get

$$G_6 = \frac{\sigma_6}{\varepsilon_6} = Q_{66} \tag{11.31}$$

Note that only four of the five constants are independent because

$$\nu_1 E_2 = \nu_2 E_1 \tag{11.32}$$

or

$$\frac{E_1}{E_2} = \frac{\nu_1}{\nu_2} \tag{11.33}$$

We can solve Eqs. (11.27)–(11.30) for the Q_{ij} to give

$$Q_{11} = \frac{E_1}{1 - \nu_1\nu_2} \qquad Q_{22} = \frac{E_2}{1 - \nu_1\nu_2} \qquad Q_{12} = \frac{\nu_1 E_2}{1 - \nu_1\nu_2} = \frac{\nu_2 E_1}{1 - \nu_1\nu_2}$$

and $Q_{66} = G_6$ is given by Eq. (11.31). Similarly, we can also show that the relationships between compliances and engineering constants are as follows:

$$S_{11} = \frac{1}{E_1} \qquad S_{22} = \frac{1}{E_2} \qquad S_{12} = -\frac{\nu_1}{E_1} = -\frac{\nu_2}{E_2} \qquad S_{66} = \frac{1}{G_6}$$

11.4 Variation of Lamina Properties with Orientation

In Sect. 11.2 we obtained the relationships between Q_{ij} and $\bar{Q}_{ij}$. It is of interest to obtain similar relationships for conventional engineering constants referred to geometric axes x–y (E_x, E_y, G_s, and ν_x) in terms of engineering constants referred to material symmetry axes 1-2 (E_1, E_2, G_6, and ν_1). Consider Eqs. (11.16) and (11.18) and let σ_x be the only nonzero stress component. Then

$$\sigma_1 = \sigma_x m^2 \tag{11.34a}$$

$$\sigma_2 = \sigma_x n^2 \tag{11.34b}$$

$$\sigma_6 = -\sigma_x mn \tag{11.34c}$$

By Hooke's law we can write for the strains in a lamina

$$\varepsilon_1 = \frac{1}{E_1}(\sigma_1 - \nu_1\sigma_2) \tag{11.35a}$$

$$\varepsilon_2 = \frac{1}{E_2}(\sigma_2 - \nu_2\sigma_1) \tag{11.35b}$$

$$\varepsilon_6 = \frac{\sigma_6}{G_6} \tag{11.35c}$$

From Eqs. (11.34)–(11.35), we get

$$\varepsilon_1 = \sigma_x\left(\frac{m^2}{E_1} - \nu_1\frac{n^2}{E_1}\right) = \sigma_x\left(\frac{m^2}{E_1} - \frac{\nu_2}{E_2}n^2\right) \tag{11.36a}$$

$$\varepsilon_2 = \sigma_x\left(\frac{n^2}{E_2} - \nu_2\frac{m^2}{E_2}\right) = \sigma_x\left(\frac{n^2}{E_2} - \frac{\nu_1}{E_1}m^2\right) \tag{11.36b}$$

$$\varepsilon_6 = -\frac{\sigma_x mn}{G_6} \tag{11.36c}$$

Since we have the strain transformation given by Eq. (11.17), we can write the inverse of Eq. (11.17) as

$$\begin{bmatrix} \varepsilon_x \\ \varepsilon_y \\ \varepsilon_s \end{bmatrix} = [T]_\varepsilon^{-1} \begin{bmatrix} \varepsilon_1 \\ \varepsilon_2 \\ \varepsilon_6 \end{bmatrix}$$

where $[T]_\varepsilon^{-1}$ can be obtained by substituting $-\theta$ for θ in Eq. (11.19). In the expanded form, we have

$$\varepsilon_x = m^2\varepsilon_1 + n^2\varepsilon_2 - mn\varepsilon_6 \tag{11.37a}$$

$$\varepsilon_y = n^2\varepsilon_1 + m^2\varepsilon_2 + mn\varepsilon_6 \tag{11.37b}$$

$$\varepsilon_s = 2(\varepsilon_1 - \varepsilon_2)mn + \varepsilon_6(m^2 - n^2) \tag{11.37c}$$

Substituting Eq. (11.36) in Eq. (11.37), we obtain

$$\varepsilon_x = \sigma_x\left[\frac{m^4}{E_1} + \frac{n^4}{E_2} + \left(\frac{1}{G_6} - \frac{2\nu_1}{E_1}\right)m^2n^2\right] \tag{11.38a}$$

$$\varepsilon_y = -\sigma_x\left[\frac{\nu_1}{E_1} - \left(\frac{1}{E_1} + \frac{2\nu_1}{E_1} + \frac{1}{E_2} - \frac{1}{G_6}\right)m^2n^2\right] \tag{11.38b}$$

$$\varepsilon_s = -\sigma_x(2mn)\left[\frac{\nu_1}{E_1} + \frac{1}{E_2} - \frac{1}{2G_6} - m^2\left(\frac{1}{E_1} + \frac{2\nu_1}{E_1} + \frac{1}{E_2} - \frac{1}{G_6}\right)\right] \tag{11.38c}$$

Now $E_x = \sigma_x/\varepsilon_x$ by definition. Combining this with Eq. (11.38a), we obtain

$$\frac{1}{E_x} = \frac{m^4}{E_1} + \frac{n^4}{E_2} + \left(\frac{1}{G_6} - \frac{2\nu_1}{E_1}\right)m^2n^2 \tag{11.39}$$

E_y can be obtained from E_x by substituting $\theta + 90°$ for θ in Eq. (11.39):

$$\frac{1}{E_y} = \frac{n^4}{E_1} + \frac{m^4}{E_2} + \left(\frac{1}{G_6} - \frac{2\nu_1}{E_1}\right)m^2n^2 \tag{11.40}$$

where $v_x = -\varepsilon_y/\varepsilon_x$ when σ_x is the applied stress. Then from Eqs. (11.38b) and (11.39), we obtain

$$\frac{v_x}{E_x} = -\frac{\varepsilon_y}{E_x\varepsilon_x} = -\frac{\varepsilon_y}{\sigma_x} = \frac{v_1}{E_1} - \left(\frac{1}{E_1} + \frac{2v_1}{E_1} + \frac{1}{E_2} - \frac{1}{G_6}\right)m^2n^2$$

or

$$v_x = E_x\left[\frac{v_1}{E_1} - \left(\frac{1}{E_1} + \frac{2v_1}{E_1} + \frac{1}{E_2} - \frac{1}{G_6}\right)m^2n^2\right] \tag{11.41}$$

Similarly, it can be shown that

$$v_y = E_y\left[\frac{v_2}{E_2} - \left(\frac{1}{E_1} + \frac{1}{E_2} + \frac{2v_1}{E_1} - \frac{1}{G_6}\right)m^2n^2\right] \tag{11.42}$$

Taking σ_s to be the only nonzero stress component, noting that $\varepsilon_s = \sigma_s/G_s$, and applying Hooke's law, we obtain

$$\frac{1}{G_s} = \frac{1}{G_6} + 4m^2n^2\left(\frac{1+v_1}{E_1} + \frac{1+v_2}{E_2} - \frac{1}{G_6}\right) \tag{11.43}$$

Figure 11.4 shows the variations of E_x, E_y, G_s, v_x, and v_y with fiber orientation θ for a 50% V_f carbon/epoxy composite. The other relevant data used in Fig. 11.4 are $E_1 = 240$ GPa, $E_2 = 8$ GPa, $G_6 = 6$ GPa, and $v_1 = 0.26$.

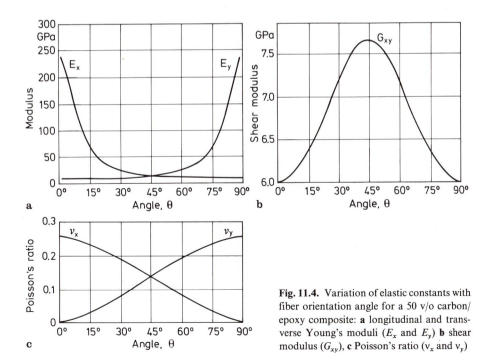

Fig. 11.4. Variation of elastic constants with fiber orientation angle for a 50 v/o carbon/epoxy composite: **a** longitudinal and transverse Young's moduli (E_x and E_y) **b** shear modulus (G_{xy}), **c** Poisson's ratio (v_x and v_y)

11.5 Analysis of Laminated Composites [5–8]

Now that we have discussed the analysis of an individual lamina, we proceed to discuss the macroscopic analysis of laminated composites. In this analysis the individual identities of fiber and matrix are ignored. Each individual lamina is treated as a homogeneous, orthotropic sheet and the laminated composite is analyzed using the classical theory of laminated plates.

It would be in order at this point to describe the way a multidirectional laminate can be defined by using a laminate code to designate the stacking sequence of laminae. Figure 11.5 shows an example of a stacking sequence. Such a stacking sequence can be described by the following code:

$$[0_2/90_2/-45_3/45_3]_s$$

This code says that starting from the bottom of the laminate, that is, at $z = -h/2$, we have a group of two plies at $0°$ orientation; then two plies at $90°$ orientation; followed by a group of three plies at $-45°$ orientation; and last, a group of three plies at $+45°$ orientation. The subscript s indicates that the laminate is symmetrical with respect to the midplane ($z = 0$); that is, the top half of the laminate is a mirror image of the bottom half. It is not necessary that a laminate composite be symmetrical. If the top half has the sequence opposite to that of the bottom half, then we shall have an asymmetric laminate. In any event, we may represent the total stacking sequence of the laminate shown in Fig. 11.5 in the following way:

$$[0_2/90_2/-45_3/45_6/-45_3/90_2/0_2]_T$$

where the subscript T indicates that the code represents the whole of the laminate thickness. Note that we have merged the two middle groups of the same ply orientation into one group. If the laminate composite consists of an odd number of laminae, the midplane will lie in the central ply.

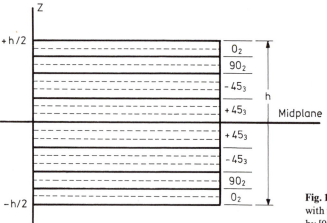

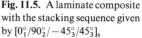

Fig. 11.5. A laminate composite with the stacking sequence given by $[0_1°/90_2°/-45_3°/45_3°]_s$

11.5.1 Basic Assumptions

We assume that the laminate thickness is small compared to its lateral dimensions. Therefore, stresses acting on the interlaminar planes in the interior of the laminate, that is, away from the free edges, are negligibly small (we shall see later that the situation is different at the free edges). We also assume that there exists a perfect bond between any two laminae. That being so, the laminae are not capable of sliding over each other and we have continuous displacements across the bond. We make yet another important assumption: namely, a line originally straight and perpendicular to the laminate midplane remains so after deformation. Actually, this follows from the perfect bond assumption which does not allow sliding between the laminae.

Finally, we have the so-called Kirchhoff assumption which states that in-plane displacements are linear functions of the thickness, and therefore the interlaminar shear strains, ε_{xz} and ε_{yz}, are negligible. With these assumptions we can reduce the laminate behavior to a two-dimensional analysis of the laminate midplane. We have the following strain–displacement relationships:

$$\varepsilon_x = \frac{\partial u}{\partial x} \qquad \varepsilon_{xy} = \frac{\partial u}{\partial y} + \frac{\partial v}{\partial x}$$

$$\varepsilon_y = \frac{\partial v}{\partial y} \qquad \varepsilon_{xz} = \frac{\partial u}{\partial z} + \frac{\partial w}{\partial x} \tag{11.44}$$

$$\varepsilon_z = \frac{\partial w}{\partial z} \qquad \varepsilon_{yz} = \frac{\partial v}{\partial z} + \frac{\partial w}{\partial y}$$

Here, u, v, and w are the displacements in the x, y, and z directions, respectively. For $i \neq j$, the ε_{ij} represent engineering shear strain components equal to twice the tensorial shear components. As per Kirchhoff's assumption, the in-plane displacements are linear functions of the thickness coordinate z. Then

$$u = u_0(x, y) + zF_1(x, y) \qquad v = v_0(x, y) + zF_2(x, y) \tag{11.45}$$

where u_0 and v_0 are displacements of the midplane. It also follows from Kirchhoff's assumptions that interlaminar shear strains ε_{xz} and ε_{yz} are zero. Therefore, from Eqs. (11.44) and (11.45) we obtain

$$\varepsilon_{xz} = F_1(x, y) + \frac{\partial w}{\partial x} = 0$$

$$\varepsilon_{yz} = F_2(x, y) + \frac{\partial w}{\partial y} = 0$$

It follows therefore that

$$F_1(x, y) = -\frac{\partial w}{\partial x} \quad \text{and} \quad F_2(x, y) = -\frac{\partial w}{\partial y} \tag{11.46}$$

The strain in the thickness direction, ε_z, is negligible, thus we can write

$$w = w(x, y)$$

That is, the vertical displacement of any point does not change in the thickness direction.

Substituting Eq. (11.46) into Eq. (11.45), we obtain

$$\varepsilon_x = \frac{\partial u}{\partial x} = \frac{\partial u_0}{\partial x} - z\frac{\partial^2 w}{\partial x^2} = \varepsilon_x^0 + zK_x \qquad (11.47a)$$

$$\varepsilon_y = \frac{\partial v}{\partial y} = \frac{\partial v_0}{\partial y} - z\frac{\partial^2 w}{\partial y^2} = \varepsilon_y^0 + zK_y \qquad (11.47b)$$

$$\varepsilon_{xy} = \frac{\partial u}{\partial y} + \frac{\partial v}{\partial x} = \frac{\partial u_0}{\partial y} + \frac{\partial v_0}{\partial x} - 2z\frac{\partial^2 w}{\partial x\partial y} = \varepsilon_{xy}^0 + zK_{xy} \qquad (11.47c)$$

Denoting ε_{xy} by ε_s and K_{xy} by K_s, as per our notation, we can rewrite the expression for ε_{xy} as

$$\varepsilon_s = \varepsilon_s^0 + zK_s \qquad (11.47d)$$

Here, ε_x^0, ε_y^0, and ε_s^0 are the midplane strains, while K_x, K_y, and K_s are the plate curvatures. We can represent these quantities in a compact form as follows:

$$\begin{bmatrix} \varepsilon_x^0 \\ \varepsilon_y^0 \\ \varepsilon_s^0 \end{bmatrix} = \begin{bmatrix} \partial u_0/\partial x \\ \partial u_0/\partial y \\ \partial u_0/\partial y + \partial v_0/\partial x \end{bmatrix} \qquad (11.48)$$

and

$$\begin{bmatrix} K_x \\ K_y \\ K_s \end{bmatrix} = - \begin{bmatrix} \partial^2 w/\partial x^2 \\ \partial^2 w/\partial y^2 \\ 2\partial^2 w/\partial x\,\partial y \end{bmatrix} \qquad (11.49)$$

Equation (11.47) can be put into the following form:

$$\begin{bmatrix} \varepsilon_x \\ \varepsilon_y \\ \varepsilon_s \end{bmatrix} = \begin{bmatrix} \varepsilon_x^0 \\ \varepsilon_y^0 \\ \varepsilon_s^0 \end{bmatrix} + z \begin{bmatrix} K_x \\ K_y \\ K_s \end{bmatrix} \qquad (11.50)$$

11.5.2 Constitutive Relationships for Laminated Composites

Consider a composite made of n stacked layers or plies; see Fig. 11.6. Let h be the thickness of the laminated composite. Then we can write, for the kth layer, the following constitutive relationship:

$$[\sigma]_k = [\bar{Q}]_k[\varepsilon]_k \qquad (11.51)$$

From the theory of laminated plates, we have the strain–displacement relationships given by Eq. (11.50). We can rewrite Eq. (11.48) as

$$[\varepsilon] = [\varepsilon^0] + z[K] \qquad (11.52)$$

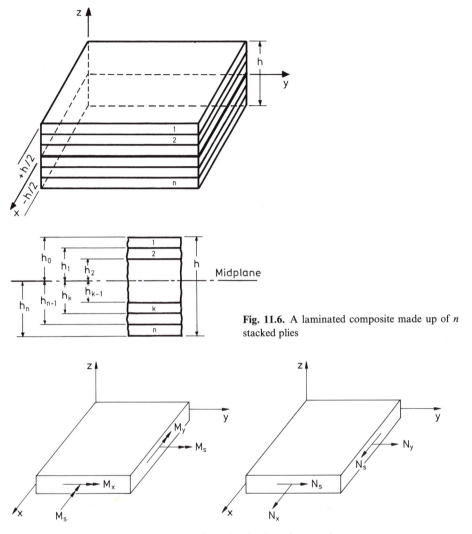

Fig. 11.6. A laminated composite made up of n stacked plies

Fig. 11.7. Force (N) and moment (M) resultants in a laminated composite

Substituting Eq. (11.52) in Eq. (11.51), we get for the kth ply

$$[\sigma]_k = [\bar{Q}]_k[\varepsilon^0] + z[\bar{Q}]_k[K] \tag{11.53}$$

Since the stresses in a laminated composite vary from ply to ply, it is convenient to define laminate force and moment resultants as shown in Fig. 11.7. These resultants of stresses and moments acting on a laminate cross section, defined as follows, provide us with a statically equivalent system of forces and moments acting at the midplane of the laminated composite. In the most general case, such a composite will have σ_x, σ_y, σ_z, σ_{xy}, σ_{yz}, and σ_{zx} as the stress components. Our laminated composite, however, is in a state of plane stress. Thus, we shall have only σ_x, σ_y, and $\sigma_{xy}(=\sigma_s)$. We define the three corresponding stress resultants as

$$N_x = \int_{-h/2}^{h/2} \sigma_x \, dz$$

$$N_y = \int_{-h/2}^{h/2} \sigma_y \, dz \tag{11.54}$$

$$N_s = \int_{-h/2}^{h/2} \sigma_s \, dz$$

These stress resultants have the dimensions of force per unit length and are positive in the same direction as the corresponding stress components. These resultants give the total force per unit length acting at the midplane. Additionally, moments are applied at the midplane which are equivalent to the moments produced by the stresses with respect to the midplane. We define the moment resultants as

$$M_x = \int_{-h/2}^{h/2} \sigma_x z \, dz$$

$$M_y = \int_{-h/2}^{h/2} \sigma_y z \, dz \tag{11.55}$$

$$M_{xy} = M_s = \int_{-h/2}^{h/2} \sigma_s z \, dz$$

This system of three stress resultants [Eq. (11.54)] and three moment resultants [Eq. (11.55)] is statically equivalent to actual stress distribution through the thickness of the composite laminate.

From Eqs. (11.52) and (11.53), we can write for the stress resultants a summation over the n plies:

$$
\begin{bmatrix} N_x \\ N_y \\ N_s \end{bmatrix} = \sum_{k=1}^{n} \int_{h_{k-1}}^{h_k} \begin{bmatrix} \sigma_x \\ \sigma_y \\ \sigma_s \end{bmatrix}_k dz
$$

$$
= \sum_{k=1}^{n} \left(\int_{h_{k-1}}^{h_k} \begin{bmatrix} \bar{Q}_{11} & \bar{Q}_{12} & \bar{Q}_{16} \\ \bar{Q}_{12} & \bar{Q}_{22} & \bar{Q}_{26} \\ \bar{Q}_{16} & \bar{Q}_{26} & \bar{Q}_{66} \end{bmatrix} \begin{bmatrix} \varepsilon_x^0 \\ \varepsilon_y^0 \\ \varepsilon_s^0 \end{bmatrix} dz \right.
$$

$$
\left. + \int_{h_{k-1}}^{h_k} \begin{bmatrix} \bar{Q}_{11} & \bar{Q}_{12} & \bar{Q}_{16} \\ \bar{Q}_{12} & \bar{Q}_{22} & \bar{Q}_{26} \\ \bar{Q}_{16} & \bar{Q}_{26} & \bar{Q}_{66} \end{bmatrix} \begin{bmatrix} K_x \\ K_y \\ K_s \end{bmatrix} z \, dz \right) \tag{11.56}
$$

Note that $[\varepsilon^0]$ and $[K]$ are not functions of z and in a given ply $[\bar{Q}]$ is not a function of z. Thus, we can simplify the above expression to

$$
\begin{bmatrix} N_x \\ N_y \\ N_s \end{bmatrix} = \sum_{k=1}^{n} \left(\begin{bmatrix} \bar{Q}_{11} & \bar{Q}_{12} & \bar{Q}_{16} \\ \bar{Q}_{12} & \bar{Q}_{22} & \bar{Q}_{26} \\ \bar{Q}_{16} & \bar{Q}_{26} & \bar{Q}_{66} \end{bmatrix} \begin{bmatrix} \varepsilon_x^0 \\ \varepsilon_y^0 \\ \varepsilon_s^0 \end{bmatrix} \int_{h_{k-1}}^{h_k} dz \right.
$$

$$
\left. + \begin{bmatrix} \bar{Q}_{11} & \bar{Q}_{12} & \bar{Q}_{16} \\ \bar{Q}_{12} & \bar{Q}_{22} & \bar{Q}_{26} \\ \bar{Q}_{16} & \bar{Q}_{26} & \bar{Q}_{66} \end{bmatrix} \begin{bmatrix} K_x \\ K_y \\ K_s \end{bmatrix} \int_{h_{k-1}}^{h_k} z \, dz \right) \tag{11.57}
$$

We can rewrite Eq. (11.57) as

$$
\begin{bmatrix} N_x \\ N_y \\ N_s \end{bmatrix} = \begin{bmatrix} A_{11} & A_{12} & A_{16} \\ A_{12} & A_{22} & A_{26} \\ A_{16} & A_{26} & A_{66} \end{bmatrix} \begin{bmatrix} \varepsilon_x^0 \\ \varepsilon_y^0 \\ \varepsilon_s^0 \end{bmatrix} + \begin{bmatrix} B_{11} & B_{12} & B_{16} \\ B_{22} & B_{22} & B_{26} \\ B_{16} & B_{26} & B_{66} \end{bmatrix} \begin{bmatrix} K_x \\ K_y \\ K_s \end{bmatrix}
\tag{11.58}
$$

or

$$
[N] = [A][\varepsilon^0] + [B][K]
\tag{11.59}
$$

where

$$
A_{ij} = \sum_{k=1}^{n} (\bar{Q}_{ij})_k (h_k - h_{k-1})
\tag{11.60}
$$

and

$$
B_{ij} = \frac{1}{2} \sum_{k=1}^{n} (\bar{Q}_{ij})_k (h_k^2 - h_{k-1}^2)
\tag{11.61}
$$

Similarly, from Eqs. (11.52) and (11.55), we can write for the moment resultants

$$
\begin{bmatrix} M_x \\ M_y \\ M_s \end{bmatrix} = \begin{bmatrix} B_{11} & B_{12} & B_{16} \\ B_{12} & B_{22} & B_{26} \\ B_{16} & B_{26} & B_{66} \end{bmatrix} \begin{bmatrix} \varepsilon_x^0 \\ \varepsilon_y^0 \\ \varepsilon_s^0 \end{bmatrix} + \begin{bmatrix} D_{11} & D_{12} & D_{22} \\ D_{12} & D_{16} & D_{26} \\ D_{16} & D_{26} & D_{66} \end{bmatrix} \begin{bmatrix} K_x \\ K_y \\ K_s \end{bmatrix}
\tag{11.62}
$$

or

$$
[M] = [B][\varepsilon^0] + [D][K]
\tag{11.63}
$$

where

$$
D_{ij} = \frac{1}{3} \sum_{k=1}^{n} (\bar{Q}_{ij})_k (h_k^3 - h_{k-1}^3)
\tag{11.64}
$$

and the B_{ij} are given by Eq. (11.61) above.

We may combine Eqs. (11.59) and (11.63) and write the constitutive equations for the laminate composite in a more compact form. Thus,

$$
\begin{bmatrix} N \\ \hline M \end{bmatrix} = \begin{bmatrix} A & | & B \\ \hline B & | & D \end{bmatrix} \begin{bmatrix} \varepsilon^0 \\ \hline K \end{bmatrix}
\tag{11.65}
$$

To appreciate the significance of the above expressions, let us examine the expression for N_x:

$$
N_x = A_{11}\varepsilon_x^0 + A_{12}\varepsilon_y^0 + A_{16}\varepsilon_s^0 + B_{11}K_x + B_{12}K_y + B_{16}K_s
$$

We note that the stress resultant is a function of the midplane tensile strains (ε_x^0 and ε_y^0), the midplane shear (ε_s^0), the bending curvatures (K_x and K_y), and the twisting (K_s). This is a much more complex situation than that observed in a homogeneous plate where tensile loads result in only tensile strains. In a laminated plate we have coupling between tensile and shear, tensile and bending, and tensile and twisting effects. Specifically, the terms A_{16} and A_{26} bring in the tension–shear coupling, while the terms B_{16} and B_{26} represent the tension–twisting coupling. The D_{16} and D_{26} terms in a similar expression for M_x represent flexure–twisting coupling.

Under certain conditions, the stress and moment resultants become uncoupled. It is instructive to examine the conditions under which some of these simplifications can result. The A_{ij} terms are the sum of ply $\bar{Q}_{ij}$ times the ply thickness [Eq. (1160)]. Thus, the A_{ij} will be zero if the positive contributions of some laminae are nullified by the negative contributions of others. Now the Q_{ij} terms of a ply are derived from orthotropic stiffnesses and, because of the form of transformation equations (11.25), $\bar{Q}_{11}$, $\bar{Q}_{12}$, $\bar{Q}_{22}$, and $\bar{Q}_{66}$ are always positive. This means that A_{11}, A_{12}, A_{22}, and A_{66} are always positive. $\bar{Q}_{16}$ and $\bar{Q}_{26}$, however, are zero for 0° and 90° orientations and can be positive or negative for θ between 0° and 90°. In fact, $\bar{Q}_{16}$ and $\bar{Q}_{26}$ are odd functions of θ; that is, for equal positive and negative orientations, they will be equal in magnitude but opposite in sign. In particular, $\bar{Q}_{16}$ and $\bar{Q}_{26}$ for a $+\theta$ orientation are equal to but opposite in sign to $\bar{Q}_{16}$ and $\bar{Q}_{26}$ values for a $-\theta$ orientation. Thus, if for each $+\theta$ ply, we have another identical ply of the same thickness at $-\theta$, then we shall have what is called a specially orthotropic laminate with respect to in-plane stresses and strains; that is, $A_{16} = A_{26} = 0$. Relative positions of such plies in the stacking sequence do not matter.

The B_{ij} terms are sums of terms involving $\bar{Q}_{ij}$ and differences of the square of z terms for the top (h_k) and bottom (h_{k-1}) of each ply. Thus, the B_{ij} terms are even functions of h_k, which means that they are zero if the laminate composite is symmetrical with respect to thickness. In other words, the B_{ij} are zero if we have for each ply above the midplane a ply identical in properties and orientation and at an equal distance below the midplane. Such a laminate is called a symmetric laminate and will have B_{ij} identically zero. This simplifies the constitutive equations for symmetric laminates, making them considerably easier to analyze. Additionally, because of the absence of bending–stretching coupling in symmetric laminates, they do not have the problem of warping encountered in nonsymmetric laminates and caused by in-plane forces induced by thermal contractions occurring during the curing of the resin matrix. Symmetric laminates will only experience thermal strains at the midplane but no flexure. The reader should realize that the origin of the [B] matrix lies not in the intrinsic orthotropy of the laminae, but in the heterogeneous (nonsymmetric) stacking sequence of the plies. Thus, a two-ply composite consisting of isotropic materials such as aluminum and steel will show a nonzero [B].

The bending matrix D_{ij} terms are defined in terms of $\bar{Q}_{ij}$ and the difference between h_k^3 and h_{k-1}^3. The geometrical contribution ($h_k^3 - h_{k-1}^3$) is always positive. Thus, as explained above for A_{ij}, D_{11}, D_{12}, D_{22}, and D_{66} are always positive. Recall that $\bar{Q}_{16}$ and $\bar{Q}_{26}$ are odd functions of θ. D_{16} and D_{26} are therefore zero for all plies oriented at 0° or 90° because these plies have $\bar{Q}_{16} = \bar{Q}_{26} = 0$. D_{16} and D_{26} can also be made zero if, for each ply oriented at $+\theta$ and at a given distance above the midplane, we have an identical ply at an equal distance below the midplane but oriented at $-\theta$. This follows from the property of the odd function of θ; that is $\bar{Q}_{16}(+\theta) = -\bar{Q}_{16}(-\theta)$, $\bar{Q}_{26}(+\theta) = -\bar{Q}_{26}(-\theta)$, while ($h_k^3 - h_{k-1}^3$) is the same for both plies. Note, however, that such a laminated composite does not have a midplane of symmetry; that is, $B_{ij} \neq 0$. In fact, D_{16} and D_{26} are not zero for any midplane symmetric laminate except for unidirectional laminates (0° or 90°) and crossplied laminates (0°/90°). We can make D_{16} and D_{26} arbitrarily small, however, by using a large enough number of plies stacked at $\pm\theta$. This is because the contributions of $+\theta$ plies to D_{16} and D_{26} are opposite in sign to those of $-\theta$ plies, and although their locations are different distances from the midplane, they tend to cancel each other.

Yet another simple stacking sequence is the quasi-isotropic sequence. Such a laminated composite can be made by having plies of identical properties oriented in such a way that the angle between any two adjacent layers is $2\pi/n$, where n is the number of plies. Such a laminate has $[A]$ independent of orientation in the plane. We call such a stacking sequence quasi-isotropic, since $[B]$ and $[D]$ are not necessarily isotropic.

The important results of this section can be summarized as follows:

$$[\sigma]_k = [\bar{Q}]_k [\varepsilon]_k$$

where $1 \leqslant k \leqslant n$ and $i, j = 1, 2, 6$.

$$\varepsilon_i = \varepsilon_i^0 + z K_i$$

$$N_i = \int_{-h/2}^{h/2} \sigma_i \, dz$$

$$M_i = \int_{-h/2}^{h/2} \sigma_i z \, dz$$

$$N_i = A_{ij} \varepsilon_j^0 + B_{ij} K_j$$

$$M_i = B_{ij} \varepsilon_j^0 + D_{ij} K_j$$

$$A_{ij} = \sum_{k=1}^{n} (\bar{Q}_{ij})_k (h_k - h_{k-1})$$

$$B_{ij} = \frac{1}{2} \sum_{k=1}^{n} (\bar{Q}_{ij})_k (h_k^2 - h_{k-1}^2)$$

$$D_{ij} = \frac{1}{3} \sum_{k=1}^{n} (\bar{Q}_{ij})_k (h_k^3 - h_{k-1}^3)$$

Symmetric Laminates

$$\bar{Q}(z) = \bar{Q}(-z)$$

Anti-Symmetric Laminates

$$\bar{Q}(z) = -\bar{Q}(-z)$$

11.6 Stresses and Strains in Laminate Composites

We saw in Sect. 11.5 that strains produced in a lamina under load depend on the midplane strains, plate curvatures, and distances from the midplane. Midplane strains and plate curvatures can be expressed as functions of an applied load system, that is, in terms of stress and moment resultants. We derived the general constitutive equation (11.65) for laminate composites. We can invert Eq. (11.65) partially or fully and obtain explicit expressions for $[\varepsilon^0]$ and $[K]$. We use Eqs. (11.59) and (11.63) for this purpose. Solving Eq. (11.59) for midplane strains, we obtain

$$[\varepsilon^0] = [A]^{-1}[N] - [A]^{-1}[B][K] \tag{11.66}$$

Substituting Eq. (11.66) in Eq. (11.63), we obtain

$$[M] = [B][A]^{-1}[N] - ([B][A]^{-1}[B] - [D])[K] \tag{11.67}$$

Combining Eqs. (11.66) and (11.67), we obtain a partially inverted form of the constitutive equation:

$$\begin{bmatrix} \varepsilon^0 \\ \hline M \end{bmatrix} = \begin{bmatrix} A^* & B^* \\ \hline C^* & D^* \end{bmatrix}\begin{bmatrix} N \\ \hline K \end{bmatrix} \tag{11.68}$$

where

$$
\begin{aligned}
[A^*] &= [A]^{-1} \\
[B^*] &= -[A]^{-1}[B] \\
[C^*] &= [B][A]^{-1} = -[B^*]^T \\
[D^*] &= [D] - [B][A]^{-1}[B]
\end{aligned}
\tag{11.69}
$$

From Eqs. (11.66) and (11.69), we can write

$$[\varepsilon^0] = [A^*][N] + [B^*][K] \tag{11.70}$$

$$[M] = [C^*][N] + [D^*][K] \tag{11.71}$$

From Eq. (11.71), we solve for $[K]$ and obtain

$$[K] = [D^*]^{-1}[M] - [D^*]^{-1}[C^*][N] \tag{11.72}$$

Substituting this value of $[K]$ [Eq. (11.72)] in Eq. (11.70), we obtain

$$
\begin{aligned}
[\varepsilon^0] &= [A^*][N] + [B^*]([D^*]^{-1}[M] - [D^*]^{-1}[C^*][N]) \\
&= ([A^*] - [B^*][D^*]^{-1}[C^*])[N] + [B^*][D^*]^{-1}[M]
\end{aligned}
\tag{11.73}
$$

We can combine Eqs. (11.72) and (11.70) to obtain the fully inverted form:

$$\begin{bmatrix} \varepsilon^0 \\ \hline K \end{bmatrix} = \begin{bmatrix} A' & B' \\ \hline C' & D' \end{bmatrix}\begin{bmatrix} N \\ \hline M \end{bmatrix} = \begin{bmatrix} A' & B' \\ \hline B' & D' \end{bmatrix}\begin{bmatrix} N \\ \hline M \end{bmatrix} \tag{11.74}$$

where

$$
\begin{aligned}
[A'] &= [A^*] - [B^*][D^*]^{-1}[C^*] \\
&= [A^*] + [B^*][D^*]^{-1}[B^*]^T \qquad ([C^*] = -[B^*]^T) \\
[B'] &= [B^*][D^*]^{-1} \\
[C'] &= -[D^*]^{-1}[C^*] = [D^*]^{-1}[B^*]^T = [B']^T = [B'] \\
[D'] &= [D^*]^{-1}
\end{aligned}
$$

Equations (11.65), (11.68), and (11.74) are useful forms of the laminate constitutive relationships. We note that each form involves obtaining the elastic properties of the lamina (from the $\bar{Q}_{ij}$ values for each lamina) and the ply stacking sequence (z coordinate).

11.7 Interlaminar Stresses and Edge Effects

The classical lamination theory used in Sect. 11.5 to describe the laminate composite behavior is rigorously correct for an infinite laminate composite. It turns out that the assumption of a generalized plane stress state is quite valid in the interior of the laminate, that is away from the free edges. At and near the free edges (extending a distance approximately equal to the laminate thickness) there exists, in fact, a three-dimensional state of stress. Under certain circumstances, there can be present rather large interlaminar stresses at the free edges which can lead to delamination of plies or matrix cracking at the free edges and thereby cause failure. Pipes and coworkers [9–13] have studied these aspects quite extensively and have clarified a number of issues. Pipes and Pagano [9] considered a four-ply laminate, $\pm\theta$ and thickness $4h_0$, under a uniform axial strain as shown in Fig. 11.8. They used a finite difference method to obtain the numerical results for a carbon epoxy composite system having plies at $\pm 45°$. The classical lamination theory states that in each ply there exists a state of plane stress with σ_x as the axial component and σ_{xy} as the in-plane shear stress component. As per the lamination theory, the stress components vary from layer to layer, but they are constant within each layer. This is correct for an infinitely wide laminate. It is incorrect for a finite-width laminate because the in-plane shear stress must vanish at the free edge surface. Figure 11.9 shows the stress distribution at the interface $z = h_0$ as a function of y/b, where $2b$ is the laminate width. The in-plane shear stress $\sigma_{xy}(=\sigma_s)$ converges to the value predicted by the lamination theory for $y/b < 0.5$, that is, away from the free edge. The axial stress component σ_x is also in accord with the lamination theory prediction for $y/b < 0.5$. The stress components σ_y, σ_z, and σ_{yz} increase near the free edge but they are quite small. The interlaminar shear stress σ_{xz}, however, has a very high

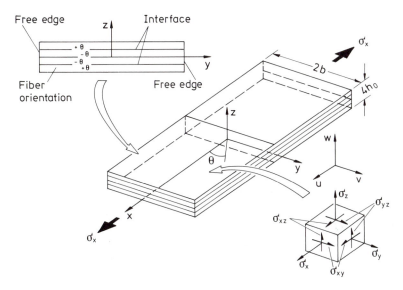

Fig. 11.8. A four-ply laminate ($\pm\theta$, thickness $4h_0$) under a uniform axial strain. (From Ref. 9, used with permission.)

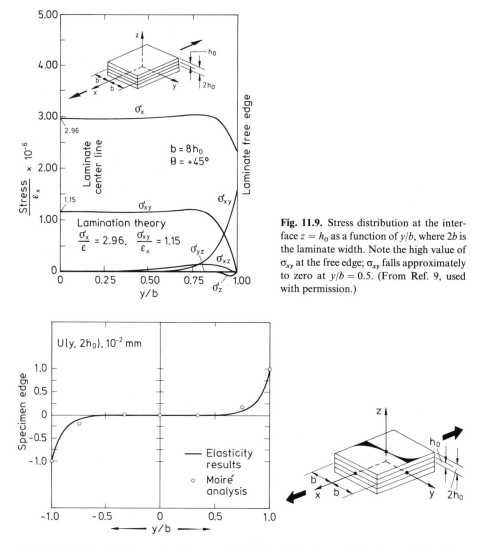

Fig. 11.9. Stress distribution at the interface $z = h_0$ as a function of y/b, where $2b$ is the laminate width. Note the high value of σ_{xy} at the free edge; σ_{xy} falls approximately to zero at $y/b = 0.5$. (From Ref. 9, used with permission.)

Fig. 11.10. Surface displacements of a symmetric angle-ply laminate subjected to axial tension. Experimental data points were determined by the Moiré technique. (From Ref. 10, used with permission.)

value at the free edge and it falls approximately to zero at $y/b = 0.5$. As can be seen from Fig. 11.9, the perturbance owing to the free edge runs through a distance approximately equal to the laminate thickness. Thus, we may regard the interlaminar stresses as a *boundary layer phenomenon* restricted to the free edge and extending inward a distance equal to the laminate thickness. Pipes and Daniel [10] confirmed these results experimentally. They used the Moiré technique to observe the surface displacements of a symmetric angle-ply laminate subjected to axial tension. Figure 11.10 shows that the agreement between experiment and theory is excellent.

An important aspect of this phenomenon of edge effects is that the laminate stacking sequence can influence the magnitude and nature of the interlaminar stresses [11–15]. It had been observed in some earlier work that identical angle-ply laminates stacked in two different sequences had different properties: the $[\pm 15°/\pm 45°]_s$ sequence had poor mechanical properties compared to the $[\pm 45°/\pm 15°]_s$ sequence. Pagano and Pipes [11] showed that interlaminar normal stress σ_z changed from tension to compression as the ply sequence was inverted. A tensile interlaminar stress at the free edge would initiate delamination there, which would account for the observed difference in the mechanical properties. Whitney [14] observed the same effect in carbon/epoxy composites in fatigue testing; namely, a specimen having a stacking sequence causing a tensile interlaminar stress at the free edge showed delaminations well before the fracture, while a specimen with a stacking sequence causing compressive interlaminar stress at the free edge showed little incidence of delaminations.

We can summarize the edge effects in laminate composites as follows:

1. The classical lamination theory of plates in plane stress is valid in the laminate interior, provided the laminate is sufficiently wide (i.e., $b/4h_0 \gg 2$).
2. Interlaminar stresses are confined to narrow regions of dimensions comparable to the laminate thickness and adjoining the free edges (i.e., $y = \pm b$).
3. The ply stacking sequence in a laminate affects the magnitude as well as the sign of the interlaminar stresses which in turn affects the mechanical performance of the laminate. Specifically, a tensile interlaminar stress at the free edge is likely to cause delaminations.

References

1   R.L. McCullough, *Concepts of Fiber–Resin Composites*, Marcel Dekker, New York, 1971, p. 16.
2   A.E.H. Love, *A Treatise on the Mathematical Theory of Elasticity*, 4th ed., Dover, New York. 1952.
3   S. Timoshenko and J.N. Goodier, *Theory of Elasticity*, McGraw-Hill, New York, 1951.
4   J.F. Nye, *Physical Properties of Crystals*, Oxford University Press, London, 1957.
5   S.W. Tsai and H.T. Hahn, *Introduction to Composite Materials*, Technomic, Westport, CT, 1980.
6   R.M. Jones, *Mechanics of Composite Materials*, Scripta Book Co., Washington, DC, 1975.
7   R.M. Christensen, *Mechanics of Composite Materials*, John Wiley & Sons, New York, 1979.
8   J.C. Halpin, *Primer on Composite Materials*, 2nd ed., Technomic, Lancaster, PA, 1984.
9   R.B. Pipes and N.J. Pagano, *J. Composite Mater.*, **4**, 538 (1970), Technomic, Lancaster, PA.
10 R.B. Pipes and I.M. Daniel, *J. Composite Mater.*, **5**, 255 (1971), Technomic, Lancaster, PA.
11 N.J. Pagano and R.B. Pipes, *J. Composite Mater.*, **5**, 50 (1971).
12 R.B. Pipes and N.J. Pagano, *J. Appl. Mech.*, **41**, 668 (1974).
13 R.B. Pipes, B.E. Kaminski, and N.J. Pagano, in *Analysis of the Test Methods for High Modulus Fibers and Composites*, ASTM STP 521, ASTM, Philadelphia, 1973, p. 218.
14 J.M. Whitney, in *Analysis of the Test Methods for High Modulus Fibers and Composites*, ASTM STP 521, ASTM, Philadelphia, 1973, p. 167.
15 D.W. Oplinger, B.S. Parker, and F.P. Chiang, *Exp. Mech.*, **14**, 747 (1974).

Suggested Reading

L.R. Calcote, *Analysis of Laminated Composite Structures*, Van Nostrand Reinhold, New York, 1969.

R.M. Christensen, *Mechanics of Composite Materials*, John Wiley & Sons, New York, 1979.

R.M. Jones, *Mechanics of Composite Materials*, Scripta Book Co., Washington, DC, 1975.

S.W. Tsai and H.T. Hahn, *Introduction to Composite Materials*, Technomic, Westport, CT, 1980.

J.R. Vinson and T.W. Chou, *Composite Materials and Their Use in Structures*, John Wiley & Sons, New York, 1975.

J.R. Vinson and R.L. Sierakowski, *The Behavior of Structures Composed of Composite Materials*, Martinus Nijhoff, Dordrecht, The Netherlands, 1986.

12. Strength, Fracture, Fatigue, and Design

12.1 Tensile Strength of Unidirectional Fiber Composites

We discussed in Chap. 10 the prediction of elastic and thermal properties, knowing the component properties. A particularly simple but crude form of this is the rule of mixtures, which works reasonably well for predicting the longitudinal elastic constants. Unfortunately, the same cannot be said for the strength of a fiber composite. It is instructive to examine why the rule of mixtures approach does not work for strength properties.

For a composite containing continuous fibers, unidirectionally aligned and loaded in the fiber direction (isostrain condition), we can write for the stress in the composite

$$\sigma_c = \sigma_f V_f + \sigma_m V_m \tag{12.1}$$

where σ the axial stress, V is the volume fraction, and the subscripts c, f, and m refer to composite, fiber, and matrix, respectively. The big question here is the value of the matrix stress σ_m. Ideally, it should be the in situ value of the matrix flow stress at a given strain. The main reason that the rule of mixtures does not work for predicting the composite strength, while it works reasonably well for Young's modulus in the longitudinal direction, is that the elastic modulus is a relatively microstructure-insensitive property, while strength is a highly microstructure-sensitive property. For example, the grain size of a polycrystalline material affects its strength but not its modulus. Thus, a response of the composite to an applied stress is nothing but the volume weighted average of the individual responses of the isolated components. Since the strength is an extremely structure-sensitive property (in the broadest sense), synergism can occur in regard to composite strength. Consider the various factors that may influence the composite strength properties. First, matrix or fiber structure may be altered during fabrication. Second, composite materials consist of two components whose thermomechanical properties are generally quite different and thus may have residual stresses and/or undergo structure alterations owing to the internal stresses. We have already discussed at length in Chap. 10 the effects of differential contraction during cooling from the fabrication temperature to ambient temperature which leads to thermal stresses large enough to deform the matrix plastically and work harden it [1–5].

Yet another source of microstructural modification of a component is a phase transformation induced by the fabrication process. In a metallic laminate composite made by roll bonding aluminum and austenitic stainless steel (type 304), it was observed that the fabrication procedure work hardened the steel as well as transformed partially the austenite to martensite [6].

We also discussed in Chap. 10 how the matrix stress state may also be influenced by rheological interaction between the components during straining [7,8]. The plastic constraint on the matrix owing to the large differences in the Poisson ratios of the matrix and the fiber, especially in the stage wherein the fiber deforms elastically while the matrix deforms plastically, can considerably alter the stress state in the composite. Thus, microstructural changes in one or both of the components, or rheological interaction between the components during straining, can lead to the phenomenon of synergism in the strength properties. In view of this, the rule mixtures should be regarded, in the best of circumstances, as an order of magnitude indicator of the strength of a composite. Nevertheless, it is instructive to consider this lower bound on the composite strength behavior. We ignore any negative deviations from the rule of mixtures caused by any fiber misalignment or the formation of a reaction product between the fiber and matrix. We also assume that the components do not interact during straining and that their properties in the composite state are the same as those in the isolated state.

We can show, following Kelly and Davies [9], that a composite must have a certain minimum fiber (continuous) volume fraction V_{min} for the composite to show a real fiber reinforcement. Assuming that the fibers are all identical and uniform, that is, all fibers have the same ultimate tensile strength, we can calculate the composite ultimate strength σ_{cu} that will be ideally attained for a strain at which fibers fracture. Thus, we can write from Eq. (12.1)

$$\sigma_{cu} = \sigma_{fu} V_f + \sigma'_m (1 - V_f) \qquad V_f > V_{min} \tag{12.2}$$

where σ_{fu} is the fiber ultimate tensile stress in the composite, σ'_m is the matrix stress at the strain corresponding to the fiber ultimate tensile strength, and $V_f + V_m = 1$. At low fiber volume fractions, a work hardened matrix can counterbalance the loss of load-carrying capacity as a result of fiber fracture. At such low V_{fs}, the matrix controls the composite strength. If all the fibers break at the same time, we must satisfy the following relationship in order to have real fiber strengthening:

$$\sigma_{cu} = \sigma_{fu} V_f + \sigma'_m (1 - V_f) \geqslant \sigma_{mu} (1 - V_f) \tag{12.3}$$

where σ_{mu} is the matrix ultimate tensile strength. The equality in this expression gives the minimum fiber volume fraction V_{min} that must be exceeded to have real reinforcement. Thus,

$$V_{min} = \frac{\sigma_{mu} - \sigma'_m}{\sigma_{fu} + \sigma_{mu} - \sigma'_m} \tag{12.4}$$

Note that the value of V_{min} increases with decreasing fiber strength.

In reality, we want the compsite strength to be more than the matrix ultimate strength in isolation. We can define a critical fiber volume fraction V_{crit} that must be exceeded for this. Thus,

$$\sigma_{cu} = \sigma_{fu} V_f + \sigma'_m (1 - V_f) \geqslant \sigma_{mu} \tag{12.5}$$

The equality in Eq. (12.5) gives

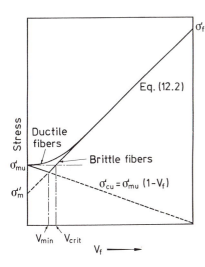

Fig. 12.1. Determination of minimum and critical fiber volume fractions for fiber reinforcement

$$V_{\text{crit}} = \frac{\sigma_{mu} - \sigma_m'}{\sigma_{fu} - \sigma_m'} \tag{12.6}$$

V_{crit} increases with increasing degree of matrix work hardening ($\sigma_{mu} - \sigma_m'$). Figure 12.1 shows graphically the determination of V_{min} and V_{crit}. Note that V_{crit} is always greater than V_{min}.

12.2 Compressive Strength of Unidirectional Fiber Composites

Fiber composites under compressive loading can be regarded, as a first approximation, as elastic columns under compression. Thus, the main failure modes in the failure of a composite are the ones that occur in the buckling of columns. Rosen [10] showed by means of photoelasticity that fiber composites fail by periodic buckling of the fibers, with the buckling wavelength being proportional to the fiber diameter. This is not surprising in view of the fact that in the analysis of a column on an elastic foundation, it is observed that the buckling wavelength depends on the column diameter. Figure 12.2 shows schematically the three situations: an unbuckled fiber composite, in-phase buckling, and out-of-phase buckling. The in-phase buckling of fibers involves shear deformation of the matrix. In such a case the composite strength in compression is proportional to the matrix shear modulus G_m; that is, $\sigma_c = G_m/V_m$, where V_m is the matrix volume fraction. For an isotropic matrix we have $G_m = E_m/2(1 + v_m)$, where E_m and v_m are the matrix Young's modulus and Poisson ratio, respectively. Thus,

$$\sigma_c = \frac{E_m}{2(1 + v_m)V_m} \tag{12.7}$$

Out-of-phase buckling of fibers involves transverse compressive and tensile strains. The compressive strength in such a case is proportional to the geometric mean of

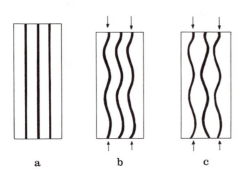

Fig. 12.2. a Unbuckled fiber composite, **b** in-phase buckling of fibers, **c** out-of-phase buckling of fibers

a   b   c

the fiber and matrix Young's moduli [10]:

$$\sigma_c = 2V_f \left(\frac{V_f E_m E_f}{3V_m} \right)^{1/2} \tag{12.8}$$

where V and E denote the volume fraction and Young's modulus, respectively, and the subscripts f and m denote fiber and matrix, respectively.

From Eqs. (12.7) and (12.8), we can see that the two failure modes in compression have a different dependence on the moduli of the components, that is,

$\sigma_c \propto (E_m E_f)^{1/2}$ out-of-phase

$\sigma_c \propto G_m$ in-phase   (12.9)

Thus, if we were to put the same fiber in two different matrices (i.e., with different matrix moduli), we should be able to distinguish between these two compressive failure modes. Lager and June [11] did just that with boron fibers in two different polymer matrices. An out-of-phase buckling mode predominated at low fiber volume fractions. At higher fiber volume fractions, fibers exerted more influence on each other and a coupled or in-phase buckling mode prevailed. The approximate nature of Eqs. (12.7) and (12.8) is easy to see. Both imply that as $V_m \rightarrow 0$ (or $V_f \rightarrow 1$), $\sigma_c \rightarrow \infty$, that is, the fibers are infinitely strong. Of course, no fibers are infinitely strong. Fiber/matrix adhesion [12] and matrix yielding [13] also affect the compressive strength of fiber composites. Piggott [14] provides a good review of the compressive properties of resins and composites as well as compressive testing of fiber composites. In the case of a laminate composite, poor interlaminar bonding results in the easy buckling of fibers. Heat and moisture as well as the ply stacking sequence can also affect the compressive strength.

12.3 Fracture Modes in Composites

A great variety of deformation modes can lead to failure of the composite. The operative failure mode depends, among other things, on loading conditions and the microstructure of a particular composite system. By microstructure we mean fiber diameter, fiber volume fraction, fiber distribution, and damage resulting from thermal stresses that may develop during fabrication and/or in service. In view of the fact that a very large number of factors can and does contribute to the fracture process in composites, it is not surprising that a multiplicity of failure modes is observed in a given composite system.

12.3.1 Single and Multiple Fracture

In general, the fiber and matrix will have different values of strain at fracture. When the component that has the smaller breaking strain fractures, for example, a brittle fiber or a brittle ceramic matrix, the load carried by this component is thrown onto the other one. If the component with a higher strain of fracture can bear this additional load, the composite will show multiple fracture of the brittle component. A manifestation of this phenomenon is fiber–bridging of the ceramic matrix (see Chap. 7). Eventually, a particular transverse section of the composite becomes so weak that the composite is unable to carry the load any further and it fails.

Consider the case of a fiber reinforced composite in which the fiber fracture strain is less than that of the matrix, for example, a ceramic fiber in a metallic matrix. The composite will then show a single fracture [12] when

$$\sigma_{fu} V_f > \sigma_{mu} V_m - \sigma'_m V_m \tag{12.10}$$

where σ'_m is the matrix stress corresponding to the fiber fracture strain and σ_{fu} and σ_{mu} are the ultimate tensile stresses of the fiber and matrix, respectively. Equation (12.10) states that when the fibers break, the matrix will not be able to support the additional load. This is commonly the case with composites containing a large quantity of brittle fibers in a ductile matrix. All the fibers break in more or less one plane and the composite also fails in that plane.

If we have a composite that satisfies the condition

$$\sigma_{fu} V_f < \sigma_{mu} V_m - \sigma'_m V_m \tag{12.11}$$

then the fibers will be broken into small segments until the matrix fracture strain is reached.

In the case where the fibers have a fracture strain greater than that of the matrix (e.g., a ceramic matrix reinforced with ductile fibers), we will have multiple fractures in the matrix. We can write the expression for this as [15]

$$\sigma_{fu} V_f > \sigma_{mu} V_m + \sigma'_f V_f \tag{12.12}$$

where σ'_f is the fiber stress corresponding to the matrix fracture strain.

12.3.2 Debonding, Fiber Pullout, and Delamination Fracture [16, 17]

Consider the stress distribution in a fiber shown in Fig. 10.13. Suppose that a crack originates in the matrix and approaches the fiber/matrix interface. In a discontinuous-fiber composite, fibers with extremities within a distance $l_c/2$ from the plane of the crack will debond and pullout of the matrix. These fibers will not break. In fact, the fraction of fibers pulling out will be l_c/l. Continuous fibers ($l > l_c$)

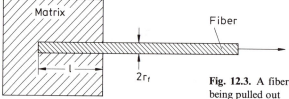

Fig. 12.3. A fiber of length l, embedded in a matrix, being pulled out

invariably have flaws distributed along their length. Thus, some of them may fracture in the plane of the crack while others may fracture away from the crack plane. This is treated in some detail further in this section. The final fracture of the composite will generally involve some fiber pullout. Consider a model composite consisting of a fiber of length l embedded in a matrix; see Fig. 12.3. If this fiber is pulled out, the adhesion between matrix and fiber will produce a shear stress τ parallel to the fiber surface. The total force acting on the fiber as a result of this stress is given by $2\pi r_f \tau l$, where r_f is the fiber radius. Let τ_i be the maximum shear stress that the interface can support and let σ_{fu} be the fiber fracture stress in tension. The maximum force caused by this normal stress on the fiber is $\pi r_f^2 \sigma_{fu}$. For maximum fiber strengthening we would like the fiber to break rather than get pulled out of the matrix. From a toughness point of view, however, fiber pullout may be more desirable. We can then write from the balance of forces the following condition for the fiber to be broken:

$$\pi r_f^2 \sigma_{fu} < 2\pi r_f \tau_i l$$

or

$$\frac{\sigma_{fu}}{4\tau_i} < \frac{l}{2r_f} = \frac{l}{d} \tag{12.13}$$

where d is the fiber diameter and the ratio l/d is the aspect ratio of the fiber.

On the other hand, for fiber pullout to occur, we can write

$$\frac{l}{d} \leqslant \frac{\sigma_{fu}}{4\tau_i} \tag{12.14}$$

The equality in this expression gives us the critical fiber length l_c for a given fiber diameter. Thus,

$$\frac{l_c}{d} = \frac{\sigma_{fu}}{4\tau_i} \tag{12.15}$$

This equation provides us with a means of obtaining the interface strength, namely, by embedding a single fiber in a matrix and measuring the load required to pull the fiber out. The load–displacement curve shows a peak corresponding to debonding, followed by an abrupt fall and wiggling about a constant stress level. Note that Eq. (12.15) gives l_c/d to be one-half of that given by Eq. (10.65). This is because in the present case the fiber is being loaded from one end only.

A point that has not been mentioned explicitly so far is that real fibers do not have uniform properties but show a statistical distribution. Weak points are

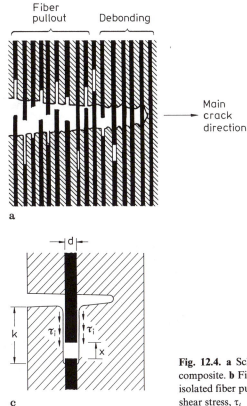

Fiber
pullout   Debonding

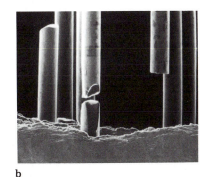

Main
crack
direction

a   b

c

Fig. 12.4. a Schematic of fiber pullout in a continuous fiber composite. **b** Fiber pullout in a B/Al system. **c** Schematic of an isolated fiber pullout through a distance x against an interfacial shear stress, τ_i

distributed along the fiber length. We treat in detail these statistical aspects of fiber strength in Sect. 12.4. Suffice it to say that it is more than likely that a fiber would break away from the main fracture plane. Interfacial debonding occurs around the fiber break point. The broken fiber parts are pulled out from their cylindrical holes in the matrix during further straining. Figure 12.4a shows schematically the fiber pullout in a continuous fiber composite, while a practical example of fiber pullout in the system B/Al is shown in Fig. 12.4b. Work is done in the debonding process as well as in fiber pullout against frictional resistance at the interface. Outwater and Murphy [16] showed that the maximum energy required for debonding is given by

$$W_d = \left(\frac{\pi d^2}{24}\right)\left(\frac{\sigma_{fu}^2}{E_f}\right)x \tag{12.16}$$

where x is the debond length.

Cottrell [18] was perhaps the first one to point out the importance of fiber pullout in regard to toughness. The length l should be large but close to l_c for maximizing the fiber pullout work and to prevent the composite from separating into two halves. It should be recognized at the same time that for $l < l_c$, the fiber will not get loaded to its maximum possible strength level and thus full fiber strengthening potential will not be realized.

We can estimate the work done in pulling out an isolated fiber in the following way, Fig. 12.4c. Let the fiber be broken a distance k below the principal crack plane, $0 < k < l_c/2$. Now let the fiber be pulled out through a distance x against an interfacial frictional shear stress, τ_i. Then the total force at that instant on the debonded fiber surface, which is opposing the pullout, is $\tau_i \pi d(k - x)$. When the fiber is further pulled out a distance dx, the work done by this force is $\tau_i \pi d(k - x)dx$. We can obtain the total work done in pulling out the fiber over the distance k by integrating. Thus,

$$\text{Work of fiber pullout} = \int_0^k \tau_i \pi d(k - x)dx = \frac{\tau_i \pi\, dk^2}{2}.$$

Now the pullout length of the fiber can vary between a minimum of 0 and a maximum of $l_c/2$. The average work of pullout per fiber is then

$$W_{fp} = \frac{1}{l_c/2} \int_0^{l_c/2} \frac{\tau_i \pi\, dk^2}{2} dk = \frac{\tau_i \pi\, dl_c^2}{24}$$

The above analysis assumes that all fibers are pulled out. In a discontinous fiber composite, it has been observed experimentally [17] that fibers with ends within a distance $l_c/2$ of the main fracture plane and that cross this fracture plane suffer pullout. Thus, it is more likely that a fraction (l_c/l) of fibers will pullout. The average work done per fiber can be written

$$W_{fp} = \left(\frac{l_c}{l}\right)\frac{\pi d\tau_i l_c^2}{24} \tag{12.17}$$

In general, fiber pullout provides a more significant contribution to composite fracture toughness than fiber/matrix debonding. The reader should appreciate the fact that debonding must precede pullout. Figure 12.5 shows schematically the variation of work of fracture with fiber length. For $l < l_c$, W_{fp} increases with fiber length because of the increasing fiber length being pulled out. W_{fp} peaks at l_c and then drops because for $l > l_c$, fiber breaks intervene and the fiber pullout decreases.

One of the attractive characteristics of composites is the possibility of obtaining an improved fracture toughness behavior together with high strength. Fracture toughness can loosely be defined as resistance to crack propagation. Consider a fibrous composite containing a crack transverse to the fibers. We can then increase the crack propagation resistance by one of the following means, each of which

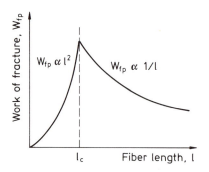

Fig. 12.5. Variation of fiber pullout work with fiber length.

involves additional work:

1. Plastic deformation of the matrix
2. Fiber pullout
3. Presence of weak interfaces, fiber/matrix separation, and deflection of the crack

For a metal matrix composite the work of fracture is the work done during plastic deformation of the matrix of failure. The work of fracture is proportional to $d(V_m/V_f)^2$ [19], where d is the fiber diameter and V_m and V_f are the matrix and fiber volume fractions, respectively. This is understandable inasmuch as in the case of large-diameter fibers, for a given V_f, the advancing crack will have to pass through a greater plastic zone of the matrix and will result in a larger work of fracture.

Fiber pullout increases the work of fracture by causing a large deformation before fracture. In this case, the controlling parameter for work of fracture is the ratio d/τ_i, where d is the fiber diameter and τ_i is the interface shear strength. In the case of discontinuous fibers, the work of fracture resulting from fiber pullout also increases with the fiber length, reaching a maximum at l_c. In the case of continuous fibers, the work to fracture increases with an increase in spacing between the defects [20, 21]. It would thus appear that one can increase the work of fracture by increasing the fiber diameter. This was discussed in Sect. 6.5 in regard to the toughness of metal matrix composites.

Crack deflection along an interface frequently follows separation of the fiber/matrix interface. This provides us with a potent mechanism of increasing the crack propagation resistance in composites. Cook and Gordon [22] have analyzed the stress distribution near a crack tip. It turns out that at the crack tip, besides a very high longitudinal stress σ_y, there also occurs a transverse stress σ_x. This transverse stress σ_x increases from zero at the crack tip to a peak value ($\sim 1/5\sigma_y$) slightly ahead of the crack tip. Thus, if there is a fiber/matrix interface in front of the crack tip such that the peak value of σ_x (acting normal to the interface) is higher than the interface strength, then the fiber/matrix interface in front of the crack tip will fail under the influence of the transverse tensile stress. The main crack would then deflect along the interface, that is, 90° from its original direction. Thus, the fiber/matrix interface can act as a crack arrestor. This is shown schematically in Fig. 12.6. This improvement in fracture toughness owing to the presence of weak interfaces has been confirmed experimentally [19, 21]. This crack deflection mechanism can be a major source of toughness in ceramic matrix composites.

Yet another related failure mode in laminate composites is the delamination fracture associated with the matrix and the fiber/matrix interface. This fracture mode is of importance in structural applications involving long-term use, for example, under fatigue conditions and where environmental effects are important.

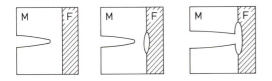

Fig. 12.6. Crack deflection at a weak interface

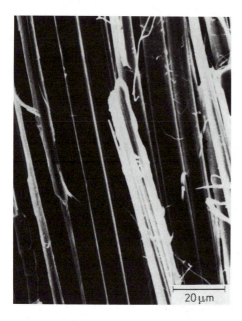

20 μm

Fig. 12.7. A delamination type fracture of Kevlar/epoxy composite. Note the characteristic fibrils. (From Ref. 7, used with permission.)

Figure 12.7 shows a delamination type fracture in a Kevlar/epoxy composite [23]. Note the presence of fibrillation which is a characteristic of the Kevlar fiber resulting from its structure as described in Chap. 2. Carbon/epoxy composites, when tested for delamination fracture, showed clean exposed fiber surfaces [23]. For details regarding this subject of delamination, the reader is referred to the proceedings of a conference on this subject [see Suggested Reading, Johnson (1985)].

12.4 Effect of Variability of Fiber Strength

Most "advanced" fibers are brittle. Thus, their strength must be characterized by a statistical distribution function. Figure 12.8a shows the strength distribution for a material showing statistical variation of strength, while Fig. 12.8b shows what is termed a Dirac delta distribution where the variability of strength is insignificant. We can safely put most metallic wires in the latter category. Most advanced fibers (B, C, SiC, Al_2O_3, etc.), however, follow some kind of statistical distribution of strength.

The Weibull statistical distribution function has been found to characterize the strength of brittle materials fairly well. For high-strength fibers also, the Weibull treatment of strength has been found to be quite adequate [24]. We follow here the treatment, due to Rosen [10, 25, 26], of this fiber strength variability problem. We can express the dependence of fiber strength on its length in terms of the following distribution function:

$$f(\sigma) = L\alpha\beta\sigma^{\beta-1} \exp(-L\alpha\sigma^\beta) \tag{12.18}$$

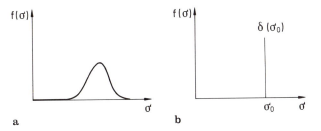

Fig. 12.8. a Strength distribution for a brittle material. **b** Strength distribution (Dirac delta) for material with insignificant variability of strength

where L is the fiber length, σ is the fiber strength, and α and β are statistical parameters. $f(\sigma)$, a probability density function, gives the probability that the fiber strength is between σ and $\sigma + d\sigma$.

We define the kth moment M_k of the statistical distribution function as

$$M_k = \int_0^\infty \sigma^k f(\sigma) d\sigma \tag{12.19}$$

Knowing that the mean strength of the fiber is given by $\bar{\sigma} = \int_0^\infty \sigma f(\sigma) d\sigma$, we can write

$$\bar{\sigma} = M_1 \tag{12.20}$$

and the standard deviation s can be expressed as

$$s = (M_2 - M_1^2)^{1/2} \tag{12.21}$$

From the Weibull distribution [Eq. (12.18)] and Eqs. (12.20) and (12.21), we obtain

$$\bar{\sigma} = (\alpha L)^{-1/\beta} \Gamma\left(1 + \frac{1}{\beta}\right) \tag{12.22}$$

and

$$s = (\alpha L)^{-1/\beta} \left[\Gamma\left(1 + \frac{2}{\beta}\right) - \Gamma^2\left(1 + \frac{1}{\beta}\right)\right]^{1/2} \tag{12.23}$$

where $\Gamma(n)$ is the gamma function given by $\int_0^\infty \exp(-x) x^{n-1} dx$. The coefficient of variation μ for this distribution then follows from

$$\mu = \frac{s}{\bar{\sigma}} = \frac{[\Gamma(1 + 2/\beta) - \Gamma^2(1 + 1/\beta)]^{1/2}}{\Gamma(1 + 1/\beta)} \tag{12.24}$$

We note that μ is a function only of the parameter β. Rosen has shown that for $0.05 \leqslant \mu \leqslant 0.5$, $\mu \simeq \beta^{-0.92}$ or $\mu \simeq 1/\beta$. In other words, β is an inverse measure of the coefficient of variation μ. For fibers characterized by the Weibull distribution [Eq. (12.18)], $\beta > 1$. For glass fibers, μ can be about 0.1, which would correspond to $\beta = 11$. For boron, μ can be between 0.2 and 0.4 and β will be between 2.7 and 5.8.

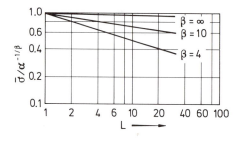

Fig. 12.9. Normalized mean fiber strength versus fiber length L

From Eq. (12.22), we can write for a unit length of fiber

$$\bar{\sigma}_1 = k\alpha^{-1/\beta} \tag{12.25}$$

where

$$k = \Gamma\left(1 + \frac{1}{\beta}\right) \tag{12.26}$$

For $\beta > 1$, we have $0.88 \leqslant k \leqslant 1.0$. Thus, we can regard the quantity $\alpha^{-1/\beta}$ as the reference strength level. We can plot Eq. (12.22) in the form of curves of $\bar{\sigma}/\alpha^{-1/\beta}$ (a normalized mean strength) against fiber length L for different β values. In Fig. 12.9, $\beta = \infty$ corresponds to a spike distribution function, that is, the Dirac delta function. In such a case, all the fibers have identical strength and there is no fiber length dependence. For $\beta = 10$ which corresponds to a $\mu \simeq 12\%$, an order of magnitude increase in fiber length produces a 20% fall in average strength. For $\beta = 4$, the corresponding fall in strength is about 50%.

If we differentiate Eq. (12.18) and equate it to zero, we obtain the statistical mode σ^*, which is the most probable strength value. Thus,

$$\sigma^* \simeq \left(\frac{\beta - 1}{\beta}\right)^{1/\beta} (\alpha L)^{-1/\beta}$$

For large β,

$$\sigma^* \simeq (\alpha L)^{-1/\beta}$$

$(\alpha L)^{-1/\beta}$, as mentioned above, is a reference stress level. The values of α and β can be obtained from experimental $\bar{\sigma}$ and μ values.

There is yet another important statistical point with regard to this variability of fiber strength. This has to do with the fact that in a unidirectionally aligned fiber composite, the fibers act in a bundle in parallel. Now the strength of a bundle of fibers whose elements do not possess a uniform strength is not the average strength of the fibers. Coleman [24] has investigated this nontrivial problem.

In the simplest case, we assume that all fibers have the same cross section and the same stress–strain curve but with different strain to fracture values. If the strength distribution function is $f(\sigma)$, the probability that a fiber will break before a certain value σ is attained is given by the cumulative strength distribution function $F(\sigma)$. We can write

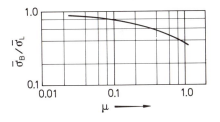

Fig. 12.10. Normalized fiber bundle strength versus variance μ of the fiber population

$$F(\sigma) = \int_0^\sigma f(\sigma)d\sigma \tag{12.27}$$

One makes a large number of measurements of strength of individual packets or bundles of fibers. Each bundle has the same large number of fibers of identical cross section and they are loaded from their extremities. From this we can find the mean fiber strength in the bundle. Daniels [27] showed that, for a very large number of fibers in the bundle, the distribution of the mean fiber strength at bundle failure σ_B attains a normal distribution, with the expectation value being given by

$$\bar{\sigma}_B = \sigma_{fu}[1 - F(\sigma_{fu})] \tag{12.28}$$

The maximum fiber strength σ_{fu} corresponds to the condition where the bundle supports the maximum load. Thus, σ_{fu} is obtained from

$$\frac{d}{d\sigma}\{\sigma[1 - F(\sigma)]\}_{\sigma=\sigma_{fu}} = 0 \tag{12.29}$$

σ_B values are characterized by the following density function for normal distribution function:

$$\omega(\sigma_B) = \frac{1}{\psi_B\sqrt{2\pi}}\exp\left[-\frac{1}{2}\left(\frac{\sigma_B - \bar{\sigma}_B}{\psi_B}\right)^2\right] \tag{12.30}$$

where ψ_B is the standard deviation given by

$$\psi_B = \sigma_{fu}\{F(\sigma_{fu})[1 - F(\sigma_{fu})]\}^{1/2}N^{-1/2} \tag{12.31}$$

where N is the number of fibers in the bundle.

As N becomes very large, not unexpectedly, the standard variation ψ_B becomes small. That is, the larger the number of fibers in the bundle, the more reproducible is the bundle strength. For bundles characterized by Eq. (12.30), we can define a cumulative distribution function $\Omega(\sigma_B)$. Thus,

$$\Omega(\sigma_B) = \int_0^{\sigma_B} \omega(\sigma_B)d\sigma_B \tag{12.32}$$

Considering the Weibull distribution [Eq. (12.18)], we have from Eq. (12.29)

$$\sigma_{fu} = (L\alpha\beta)^{-1/\beta} \tag{12.33}$$

and from Eq. (12.28), it follows that

$$\bar{\sigma}_B = (L\alpha\beta e)^{-1/\beta} \tag{12.34}$$

Comparing this mean fiber bundle strength value [Eq. (12.34)] to the mean value of the fiber strength obtained from equal-length fibers tested individually [Eq. (12.22)], we note that when there is no dispersion in fiber strength, the mean bundle strength equals the mean fiber strength, see Fig. 12.10. As the coefficient of variation of fiber strength increases above zero, the mean bundle strength decreases and, in the limit of an infinite dispersion, tends to zero. For a 10% variance, the mean bundle strength is about 80% of the mean fiber strength, while for a variance of 25%, the bundle strength is about 60% of the mean fiber strength.

In view of the statistical distribution of fiber strength, it is natural to extend these ideas to composite strength. We present here the treatment introduced by Rosen and coworkers [25, 26]. On straining a fiber composite, fibers fracture at various points before a complete failure of the composite. There occurs an accumulation of fiber fractures with increasing load. At a certain point, one transverse section will be weakened as a result of the statistical accumulation of fiber fractures, hence the name *cumulative weakening model of failure*. Since the fibers have nonuniform strength, it is expected that some fibers will break at very low stress levels. Figure 12.11 shows the perturbation in stress state when a fiber fracture occurs. At the point of fiber fracture, the tensile stress in the fiber drops to zero. From this point the tensile stress in the fiber increases along the fiber length along the two fiber segments as per the load transfer mechanism by interfacial shear described in Chap. 10. These local stresses can be very large. As a result, either the matrix would yield in shear or an interfacial failure would occur. Additionally, this local drop in stress caused by a fiber break will throw the load onto adjoining fibers, causing stress concentrations there (see Fig. 12.11). Upon continued straining, progressive fiber fractures cause a cumulative weakening and a redistribution of the load in the composite. After the first fiber fracture, the interfacial shear stresses may cause delamination between this broken fiber and the matrix as shown in Fig. 12.12a. When this happens, the broken and delaminated fiber becomes totally ineffective and the composite behaves as a bundle of fibers. The second alternative is that a crack, starting from the first fiber break, propagates through the other fibers in a direction normal to the fibers; see Fig. 12.12b. Such a situation will occur only if the

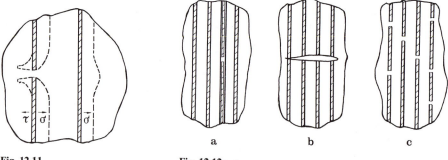

Fig. 12.11   **Fig. 12.12a–c**

Fig. 12.11. Perturbation of stress state caused by a fiber break

Fig. 12.12. Fracture models: **a** interface delamination where the composite acts as a fiber bundle, **b** first fiber fracture turns into a complete fracture, **c** cumulative damage—statistical

fiber and matrix are very strongly bonded and if the major component is very brittle. In the absence of these two modes, cumulative damage results. With increased loading, additional fiber fractures occur and a statistical distribution is obtained; see Fig. 12.12c.

Rosen [26] considers in this model that the composite strength is controlled by a statistical accumulation of failures of individual volume elements that are separated by barriers to crack propagation; thus, these elements fail independently. The load on the matrix is ignored. Increased loading leads to individual fiber breaks at loads less than the ultimate fracture load of the composite. An individual fiber break does not make the whole fiber length ineffective, it only reduces the capacity of the fiber being loaded in the vicinity of fiber failure. The stress distribution in the fiber is one of full load over its entire length less a length near the break over which the load is zero. This length is called the ineffective length. Thus, the composite is considered to be made up of a series of layers, each layer consisting of a packet of fiber elements embedded in the matrix. The length of the fiber element or the packet height is equal to the ineffective length.

As the load is increased, fiber breaks accumulate until at a critical load a packet of elements is unable to transmit the applied load and the composite fails. Thus, composite failure occurs because of this weakened section. A characteristic length δ corresponding to the packet height must be chosen. The term δ is defined as the length over which the stress attains a certain fraction ϕ of the unperturbed stress in the fiber. Rosen took ϕ equal to that length over which the stress increases elastically from zero (at a fiber end) to 90% of the unperturbed level. One derives the stress distribution in the fiber elements and packets. The theory of the weakest link is then applied to obtain the composite strength. In the case of fibers characterized by the Weibull distribution, the stress distribution in the fiber elements is

$$w(\sigma) = \delta\alpha\beta\sigma^{\beta-1}\exp(-\delta\alpha\sigma^{\beta}) \qquad (12.35)$$

where δ is the fiber element length (i.e., the ineffective length).

The composite is now a chain, the strength of whose elements is given by a normal distribution function $w(\sigma_B)$. The strength of a chain having m links of this population is characterized by a distribution function $\lambda(\sigma_c)$:

$$\lambda(\sigma_c) = mw(\sigma_c)[1 - \Omega(\sigma_c)]^{m-1} \qquad (12.36)$$

where

$$\Omega(\sigma_c) = \int_0^{\sigma_c} w(\sigma_c)d\sigma \qquad (12.37)$$

Suppose now that the number of elements N in a composite is so large that the standard deviation of the packet tends to zero. Then the statistical mode of composite strength is equal to $\bar{\sigma}_B$ [Eq. (12.28)].

In the case of a Weibull distribution, it follows from Eq. (12.34) that

$$\sigma_c^* = (\delta\alpha\beta e)^{-1/\beta} \qquad (12.38)$$

and the statistical mode of the tensile strength of the composite is

$$\sigma^* = V_f(\delta\alpha\beta e)^{-1/\beta} \qquad (12.39)$$

where V_f is the fiber volume fraction.

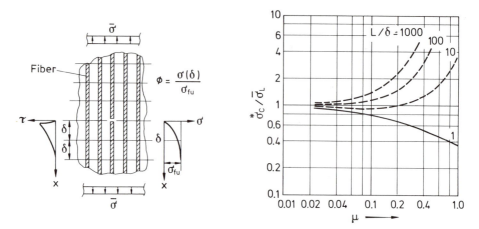

Fig. 12.13. Normalized composite strength $\sigma_c^*/\bar{\sigma}_L$ versus variance μ. (From Fiber Composite Materials, American Society for Metals, 1965, p 39, 38. With permission.)

It should be pointed out that δ will be of the order of 10–100 fiber diameters, and thus much smaller than the gage length used for individual tests.

If we compare the (cumulative) average strength of a group of fibers of length L with the expected fiber strength from Eqs. (12.24) and (12.38), we get

$$\frac{\sigma_c^*}{\bar{\sigma}_L} = \left(\frac{L}{\delta\beta e}\right)^{1/\beta} \frac{1}{\Gamma(1 + 1/\beta)} \qquad (12.40)$$

For $\beta = 5$ and $L/\delta = 100$, we have $\sigma_c^*/\bar{\sigma}_L = 1.62$; that is, the composite will be much stronger than what we expect from individual fiber tests. We can plot the composite strength σ_c^*, normalized with respect to the average strength $\bar{\sigma}_L$, of individual fibers of length L against μ, the variance of individual fiber strengths; see Fig. 12.13. The curves shown are for different values of the ratio L/δ. For $L/\delta = 1$, that is, the fiber length is equal to the ineffective length, the statistical mode of composite strength is less than the average fiber strength. This difference between the two increases with an increase in μ of the fibers. For a more realistic ratio, for example, $L/\delta > 10$, we note that the composite strength is higher than the average fiber strength.

Modification of this cumulative weakening model has been proposed by Zweben and Rosen [28] which takes into account the redistribution of stress that results at each fiber break; that is, there is greater probability that fracture will occur in immediately adjacent fibers because of a stress magnification effect.

12.5 Strength of an Orthotropic Lamina

We saw in Chapters 10 and 11 that fiber reinforced composites are anisotropic in elastic properties. This results from the fact that the fibers are a lot stiffer than the matrix and the fact that the fibers are aligned in the matrix. Fibers, however, are not only stiffer than the matrix, but they are also a lot stronger than the matrix. Therefore, not unexpectedly, fiber reinforced composites show anisotropy in strength properties as well. Quite frequently, the strength in the longitudinal direction is as much as an order of magnitude greater than that in the transverse direction.

It is of great importance for design purposes to be able to predict the strength of a composite under the loading conditions prevailing in service. The use of a failure criterion gives us information about the strength under combined stresses. We assume, for simplicity, that the material is homogeneous; that is, its properties do not change from point to point. In other words, we treat a fiber reinforced lamina as a homogeneous, orthotropic material. For a detailed account of the failure criteria and their experimental correlation, the reader is referred to a review by Rowlands [29]. We present below a brief account.

12.5.1 Maximum Stress Theory

Failure will occur when any one of the stress components is equal to or greater than its corresponding allowable or intrinsic strength. Thus, failure would occur if

$$\begin{aligned}
\sigma_1 &\geqslant X_1^T & \sigma_1 &\leqslant -X_1^C \\
\sigma_2 &\geqslant X_2^T & \sigma_2 &\leqslant -X_2^C \\
\sigma_6 &\geqslant S & \sigma_6 &\leqslant S
\end{aligned} \tag{12.41}$$

where X_1^T is the ultimate uniaxial tensile strength in the fiber direction, X_1^C is the ultimate uniaxial compressive strength in the fiber direction, X_2^T is the ultimate uniaxial tensile strength transverse to the fiber direction, X_2^C is the ultimate uniaxial compressive strength transverse to the fiber direction, and S is the ultimate planar shear strength. When any one of the inequalities indicated in Eq. (12.41) is attained, the material will fail by the failure mode related to that stress inequality. No failure mode interaction is permitted in this criterion. Consider an orthotropic lamina, that is, a unidirectional fiber reinforced prepreg subjected to a uniaxial tensile stress σ_x in a direction at an angle θ with the fiber direction. We then have for the stress components in 1-2 system

$$\begin{bmatrix} \sigma_1 \\ \sigma_2 \\ \sigma_6 \end{bmatrix} = [T]_\sigma \begin{bmatrix} \sigma_x \\ \sigma_y \\ \sigma_s \end{bmatrix} \tag{12.42}$$

where

$$[T]_\sigma = \begin{bmatrix} m^2 & n^2 & 2mn \\ n^2 & m^2 & -2mn \\ -mn & mn & m^2 - n^2 \end{bmatrix} \qquad m = \cos\theta, \; n = \sin\theta$$

Using the fact that only σ_x is nonzero, we obtain

$$\sigma_1 = \sigma_x m^2$$

$$\sigma_2 = \sigma_x n^2$$

$$\sigma_6 = \sigma_x mn$$

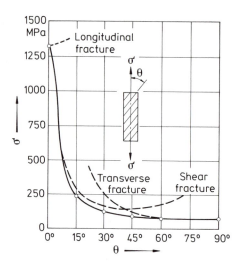

Fig. 12.14. Variation of strength with fiber orientation for boron epoxy. Quadratic interaction criterion (solid curve) shows better agreement with experimental data than the maximum stress criterion (dashed curve)

According to the maximum stress criterion, the three possible failure modes are:

$$\sigma_x = \frac{X_1^T}{m^2} \quad \text{(longitudinal tensile)}$$

$$\sigma_x = \frac{X_2^T}{n^2} \quad \text{(transverse tensile)} \tag{12.43}$$

$$\sigma_x = \frac{S}{mn} \quad \text{(planar shear)}$$

With varying θ, the failure mode will change from longitudinal tension to planar shear to transverse tension, as shown by the dashed lines in Fig. 12.14. Poor agreement with experiment, particularly around $\theta = \pi/4$, is clear from this figure. This indicates that at intermediate angles failure mode interactions do occur.

12.5.2 Maximum Strain Criterion

This criterion is analogous to the maximum stress criterion. Failure occurs when any one of the strain components is equal to or greater than its corresponding allowable strain. Thus,

$$\varepsilon_1 \geqslant e_1^T \qquad \varepsilon_1 \leqslant -e_1^C$$

$$\varepsilon_2 \geqslant e_2^T \qquad \varepsilon_2 \leqslant -e_2^C \tag{12.44}$$

$$\varepsilon_6 \geqslant e_6 \qquad \varepsilon_6 \leqslant e_6$$

where e_1^T is the ultimate tensile strain in the fiber direction, e_1^C is the ultimate compressive strain in the fiber direction, e_2^T is the ultimate tensile strain in the transverse direction, e_2^C is the ultimate compressive strain in the transverse direction, and e_6 is the ultimate planar shear strain. This criterion is also not very satisfactory.

12.5.3 Maximum Work (or the Tsai–Hill) Criterion

According to the Tsai–Hill criterion, failure of an orthotropic lamina will occur under a general stress state when

$$\frac{\sigma_1^2}{X_1^2} - \frac{\sigma_1 \sigma_2}{X_1^2} + \frac{\sigma_2^2}{X_2^2} + \frac{\sigma_6^2}{S^2} \leq 1 \tag{12.45}$$

where X_1, X_2, and S are the longitudinal tensile failure strength, the transverse tensile failure strength, and the in-plane shear failure strength, respectively. If compressive stresses are involved, then the corresponding compressive failure strengths should be used.

Consider again a uniaxial stress σ_x applied to an orthotropic lamina. Then

$$\sigma_1 = \sigma_x m^2$$
$$\sigma_2 = \sigma_x n^2 \tag{12.46}$$
$$\sigma_6 = \sigma_x mn$$

where $m = \cos\theta$ and $n = \sin\theta$. Substituting these values in Eq. (12.45), we have

$$\frac{m^4}{X_1^2} + \frac{n^4}{X_2^2} + m^2 n^2 \left(\frac{1}{S^2} - \frac{1}{X_1^2} \right) < \frac{1}{\sigma_x^2} = \frac{1}{\sigma_\theta^2} \tag{12.47}$$

12.5.4 Quadratic Interaction Criterion

As the name indicates, this criterion takes into account the stress interactions. Tsai and Wu [30] proposed this modification of the Hill theory for a lamina by adding some additional terms. Tsai and Hahn [31] provide a good account of this criterion. According to this theory, the failure surface in stress space can be described by a function of the form

$$f(\sigma) = F_i \sigma_i + F_{ij} \sigma_i \sigma_j = 1 \, i, j = 1, 2, 6 \tag{12.48}$$

where F_i and F_{ij} are the strength parameters. For the case of plane stress, $i, j = 1, 2, 6$ and we can expand Eq. (12.48) as follows:

$$F_1 \sigma_1 + F_2 \sigma_2 + F_6 \sigma_6 + F_{11} \sigma_1^2 + F_{22} \sigma_2^2 + F_{66} \sigma_6^2 + 2F_{12} \sigma_1 \sigma_2$$
$$+ 2F_{16} \sigma_1 \sigma_6 + 2F_{26} \sigma_2 \sigma_6 = 1 \tag{12.49}$$

For the orthotropic lamina, sign reversal for normal stresses, whether tensile or compressive, is important. The linear stress terms provide for this difference. For the shear stress component, the sign reversal should be immaterial. Thus, terms containing the first degree shear stress must vanish. These terms are $F_{16} \sigma_1 \sigma_6$, $F_{26} \sigma_2 \sigma_6$, and $F_6 \sigma_6$. The stress components in general are not zero. Therefore, for these three terms to vanish we must have

$$F_{16} = F_{26} = F_6 = 0$$

Equation (12.49) is now simplified to

$$F_1 \sigma_1 + F_2 \sigma_2 + F_{11} \sigma_1^2 + F_{22} \sigma_2^2 + F_{66} \sigma_6^2 + 2F_{12} \sigma_1 \sigma_2 = 1 \tag{12.50}$$

There are six strength parameters in Eq. (12.50). We can measure five of these by the following simple tests:

Longitudinal (Tensile and Compressive) Tests

If $\sigma_1 = X_1^T$, then $F_{11}(X_1^T)^2 + F_1 X_1^T = 1$.
If $\sigma_1 = -X_1^C$, then $F_{11}(X_1^C)^2 - F_1 X_1^C = 1$.
From these we get

$$F_{11} = \frac{1}{X_1^T X_1^C} \tag{12.51}$$

and

$$F_1 = \frac{1}{X_1^T} - \frac{1}{X_1^C} \tag{12.52}$$

Transverse (Tensile and Compressive) Tests

If X_2^T and X_2^C are the transverse tensile and compressive strengths, respectively, then proceeding as above, we get

$$F_{22} = \frac{1}{X_2^T X_2^C} \tag{12.53}$$

and

$$F_2 = \frac{1}{X_2^T} - \frac{1}{X_2^C} \tag{12.54}$$

Longitudinal Shear Test

If S is the shear strength, we have

$$F_{66} = \frac{1}{S^2} \tag{12.55}$$

Thus, we can express all the failure constants except F_{12} in terms of the ultimate intrinsic strength properties. F_{12} is the only remaining parameter and it must be evaluated by means of a biaxial test, not a small inconvenience. Many workers [32, 33] have proposed variations of the Tsai–Wu criterion involving F_{12} explicitly in terms of uniaxial strengths. Tsai and Hahn [31] suggest that, in the absence of other data, $F_{12} = -0.5\sqrt{F_{11}F_{22}}$. It turns out, however, that small changes in F_{12} can significantly affect the predicted strength [31]. Figure 12.14 shows for the boron/epoxy system the variation of strength with orientation assuming $F_{12} = 0$ [34]. The intrinsic properties of this system are as follows:

$$X_1^T = 27.3 \text{ MPa} \qquad X_2^T = 1.3 \text{ MPa} \qquad S = 1.4 \text{ MPa}$$

$$X_1^C = 52.4 \text{ MPa} \qquad X_2^C = 6.5 \text{ MPa}$$

Note the excellent agreement between the curve computed using the quadratic interaction criterion and the experiment. The agreement with the maximum stress criterion (dashed curve) is poor.

12.6 Fatigue of Laminate Composites

Just as in the case of static failure, fatigue failure mechanisms in fiber reinforced composites are quite different from those in monolithic, homogeneous
materials such as metals. Hahn and Lorenzo [35] have reviewed the fatigue failure
mechanisms in composite laminates. Fatigue failure in metals, for example, occurs
as a result of initiation and growth of a principal crack. Fiber reinforced laminate
composites, on the other hand, can sustain a variety of subcritical damage (matrix
cracking, fiber fracture, fiber/matrix decohesion, ply cracking, and delamination).
By ply cracking we mean mostly cracking in the weak phase, generally matrix, and
along the fiber/matrix interface with little or no fiber breakage. Ply cracking results
in a relaxation of stress in that ply and further cracking in that ply becomes difficult
with continued cycling. The existing cracks may grow into interfaces, causing ply
delamination and a reduction of the stress concentration on the neighboring plies.
The composite laminate fails when the strong fibers in the loading direction fail. A
comparison of damage accumulation in composite laminates and homogeneous
materials under constant-stress-amplitude fatigue is shown schematically in Fig.
12.15 [35]. By damage we mean crack density in composites and crack length in
homogeneous materials. The damage ratio is current damage normalized with
respect to damage at final failure. Note that the damage in laminates accelerates
and then decelerates with cycling, while it accelerates monotonically in homogeneous materials.

The various forms of damage mentioned above result in a reduction of the
load-carrying capacity of the laminate composite, which in turn manifests itself
as a reduction of laminate stiffness and strength [36–41]. Figure 12.16 depicts
schematically the changes in crack density, delamination, and modulus in composite laminates under fatigue [35]. Reifsnider et al. [41] have modeled the fatigue
development in laminate composites as occurring in two stages. In the first stage,
homogeneous, noninteractive cracks appear in individual plies. In the second stage,
the damage gets localized in zones of increasing crack interaction. The transition
from stage one to stage two occurs at what has been called as the *characteristic
damage state* (CDS), which consists of a well-defined crack pattern characterizing
saturation of the noninteractive cracking. Talreja [40] has used this model to
determine the probability distribution of the residual strength and the probability
distribution of the number of cycles required to attain the CDS. A number of

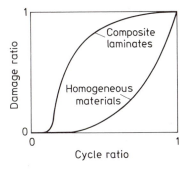

Fig. 12.15. Comparison of damage accumulation in
composite laminates and homogeneous materials under
constant-stress-amplitude fatigue. (Reprinted with permission from H.T. Hahn and L. Lorenzo, in Advances in
Fracture Research, ICF6, copyright (1984), Pergamon
Press.)

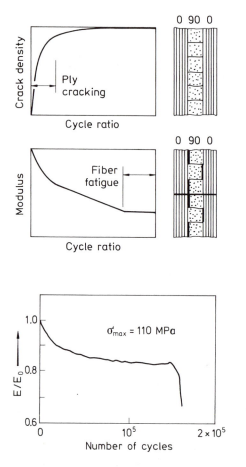

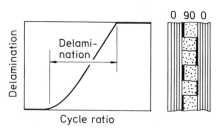

Fig. 12.16. Schematic of changes in crack density, delamination, and modulus in composite laminates under fatigue. (Reprinted with permission from H.T. Hahn and L. Lorenzo, in Advances in Fracture Research, ICF6, copyright (1984), Pergamon Press.)

Fig. 12.17. Stiffness reduction curve for a $(0°/90°)_s$ glass fiber PMC. (From Ref. 44, used with permission.)

researchers have experimentally related the stiffness changes to the accumulated damage [38, 40, 42–45]. It can safely be said that the change in stiffness values is a good indicator of the damage in composites. An actual stiffness reduction curve for a $[0°/90°]_s$ glass fiber reinforced laminate, taken from the work of Ogin et al. [44], is shown in Fig. 12.17. The modulus reduction rate $(-1/E_0)dE/dN$, at a given value of E/E_0, is the tangent to this curve at that value of E/E_0 where E is the current modulus, E_0 is the initial modulus of the uncracked material, and N is the number of cycles; see Fig. 12.18. This rate of modulus reduction was well described over the range of peak fatigue stress σ_{max} between 110 and 225 MPa by the expression

$$-\frac{1}{E_0}\frac{dE}{dN} = A\left(\frac{\sigma_{max}^2}{E_0^2(1 - E/E_0)}\right)^n$$

where A and n are constants. Ogin et al. [44] integrated this equation to obtain a diagram relating modulus reduction to number of cycles for different stress levels; see Fig. 12.19. Such diagrams have obvious design utility.

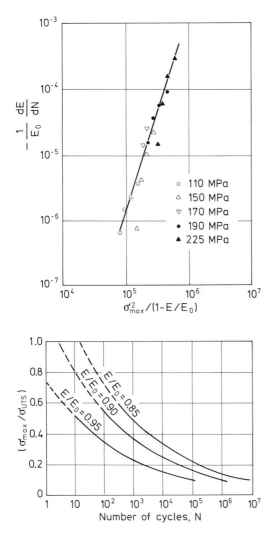

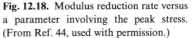

Fig. 12.18. Modulus reduction rate versus a parameter involving the peak stress. (From Ref. 44, used with permission.)

Fig. 12.19. Number of cycles required to attain a given stiffness reduction when cycled at different fractions of the ultimate tensile strength σ_{UTS}. (From Ref. 44, used with permission.)

12.7 Designing with Composite Materials

Composite materials, particularly fiber reinforced composite materials, are not just another kind of new material. When designing with composites, one must take into account their special characteristics. First, composite materials are inherently heterogeneous at a microstructural level, consisting as they do of two components having different elastic moduli, different strengths, different thermal expansion coefficients, and so on. We saw in the micromechanical analysis (Chap. 10) that the structural and physical properties of composites are functions of (a) component characteristics, (b) geometric arrangement of one component in the other, and (c) interface characteristics. Even after selecting the two basic components, one can obtain a range of properties by manipulating the items listed in (b) and (c).

Second, the conventional, monolithic materials generally are quite isotropic; that is, their properties do not show any marked preference for any particular direction, whereas fiber composites are highly anisotropic because of their very nature. The analysis and design of composites should take into account this strong directionality of properties, or, rather, this anisotropy of fiber composites must be exploited to the fullest advantage by the designer. Figures 10.4, 12.14, and 10.7 show the marked influence of fiber orientation on elastic moduli, strength, and thermal expansion, respectively. For conventional materials, the designer needs only to consult a handbook or a manual to obtain one unambiguous value of, say, modulus or any other property. For composite materials, however, the designer has to consult what are called "performance charts" or "carpet plots" representing a particular property of a given composite system. Figure 12.20 shows such plots for Young's modulus and the tensile strength, both in the longitudinal direction, for a 65% V_f carbon fiber/epoxy composite having a $0°/\pm45°/90°$ ply sequence [46]. A conventional material, say aluminum, would be represented by just one point on such graphs. The important and distinctive point is that, depending on the components, their relative amounts, and the ply stacking sequence, we can obtain a range of properties in fiber composites. In other words, one can tailor-make a composite material as per the final objective.

A good example of the versatility and flexibility of composites can be had from Fig. 12.20. Say our material specifications require, for an application, a material that is stronger than steel in the x direction ($\sigma_x = 500$ MPa) and a stiffness in the y direction equal to that of aluminum ($E = 70$ GPa). We can then pick a material combination that gives us a composite with these characteristics. Using Fig. 12.20a and b, we can choose the following ply distribution:

60% at 0°

20% at 90°

20% at ±45°

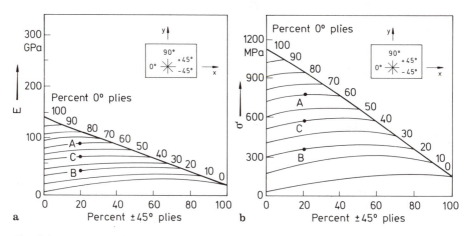

Fig. 12.20. Carpet plots for 65% V_f carbon/epoxy with a $0°/\pm45°/90°$ stacking sequence: **a** Young's modulus, **b** tensile strength. (From Ref. 46, used with permission.)

Point A has a $\sigma_x = 725$ MPa and an $E_x = 90$ GPa. If we interchange the x and y coordinates and the amounts of 0° and 90° plies, we get point B, which corresponds to a $\sigma_y = 340$ MPa and an $E_y = 40$ GPa. Thus, σ_x is now higher than required and E_y is too low. We now take some material from 0° and move it to 90°. Consider the following ply distribution:

40% at 0°

40% at 90°

20% at ±45°

This is represented by point C, having $\sigma_x = 580$ MPa and $E_y = 70$ GPa. Reversing the x and y coordinates gives again point C with $\sigma_x = 580$ MPa and $E_y = 70$ GPa. Point C then meets our initial specifications. Stacking sequences other than 0°/±45°/90° can also satisfy one's requirements. A 0°/±30°/±60°/90° arrangement gives quasi-isotropic properties in the plane. Carpet plots, however, work for three-ply combinations and the sequence 0°/±45°/90° has some other advantages as described below.

It should be pointed out that carpet plots for strengths in compression and shear, shear modulus, and thermal expansion coefficients can also be made. Figure 12.21 compares the thermal expansion coefficients of some metals and composites as a function of fiber orientation [47]. Both Kevlar and carbon fibers have negative expansion coefficients parallel to the fiber axis. This interesting characteristic allows for making composites with well-controlled expansion characteristics. Although aerospace applications of composite materials have been very much in the forefront, one should appreciate the fact that composites, in one form or another, have been in use in the electrical and electronic industry for quite a long time. These applications range from cables to printed circuit boards. A good review of applications of composites in the electronic packaging industry has been provided by Seraphim et al. [48]. An interesting example is that of the leadless chip carrier (LCC) that allows electronic designers to pack more electrical connections into less space than is possible with the conventional flat packs or dual in-line

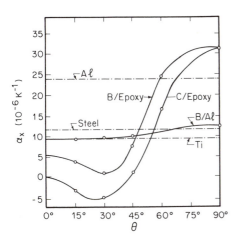

Fig. 12.21. Variation of thermal expansion coefficients with orientation θ. (From Composite Materials in Engineering Design, American Society for Metals, 1973, p 335. With permission.)

packages, which involve E-glass/epoxy resin laminates. Glass/epoxy laminate, however, is not very suitable for LCC as a substrate because it expands and contracts much more than the alumina chip carrier (α of E-glass/epoxy varies in the range of 14–$17 \times 10^{-6} \text{ K}^{-1}$, while that of alumina is $\sim 6 \times 10^{-6} \text{ K}^{-1}$). Kevlar ($40$–$50$ v/o)/epoxy composite provides the desired matching expansion characteristics with those of the alumina chip carrier. Thus, the thermal fatigue problem associated with the E-glass/epoxy composites and the ceramic chip carrier is avoided.

We saw in Chap. 11 that laminate composites show coupling phenomena—tension–shear, tension–bending, and bending–twisting coupling. We also saw in Chap. 10 that certain special ply sequences can simplify the analyses. Thus, using a balanced arrangement (for every $+\theta$ ply there is a $-\theta$ ply in the laminate), we can eliminate shear coupling. In such an arrangement, the shear distortion of one layer is compensated by an equal and opposite shear distortion of the other layer. Yet another simplifying arrangement consists of stacking the plies in a symmetrical manner with respect to the laminate midplane. Such a symmetrical laminate gives $[B_{ij}]$ identically zero and thus eliminates stretching–bending and bending–torsion coupling.

The phenomenon of edge effects described in Sect. 11.7 should also be taken into account while arriving at a ply stacking sequence. Because of the edge effects, individual layers deforming differently under tension give rise to out-of-plane shear and bending in the neighborhood of the free edges. An arrangement that gives rise to compressive stresses in the thickness direction in the vicinity of the edges is to be preferred over one that gives rise to tensile stresses. The latter would tend to cause undesirable delamination in the composite. Figure 12.22 shows delaminations as observed by contrast-enhanced X-ray radiography, in a $[0°/\pm45°/90°]_s$ carbon/epoxy laminate upon fatigue testing [49]. The delaminations proceed from both sides toward the specimen center. The $[90°/\pm45°/0°]_s$ laminate showed relatively less damage than the $[0°/\pm45°/90°]_s$ laminate. Figure 12.23 compares the crack patterns at the free edges of $[0°/\pm45°/90°]_s$ and $[90°/\pm45°/0°]_s$ laminates shortly before failure [49]. Note the greater degree of delaminations in $[0°/\pm45°/90°]_s$ than in $[90°/\pm45°/0°]_s$.

A little reflection will convince the reader that extensive use of computers can be made in the design and analyses of laminate fiber composites. In view of the rather tedious matrix calculations involved, this should not be surprising at all. Researchers at the U.S. National Aeronautics and Space Administration (NASA) Lewis Research Center have been in the forefront in developing computer codes to analyze and design fiber composite structures in a cost effective manner. Murthy and Chamis [50] describe one such code called *integrated composites analyzer* (ICAN). According to the authors, ICAN "is a synergistic combination" of some earlier computer codes. It uses micromechanics and the laminate theory to provide a "comprehensive analysis/design capability for structural composites." It should be pointed out that such codes deal with polymer matrix laminated fiber composites. ICAN, for example, has its own database of material properties of commonly used fibers and resin matrices. Among other things, the code specifies the interply layers, laminate failure stresses, free edge stresses, and probable delamination locations around a hole. The reader is referred to Ref. 51 for details of ICAN.

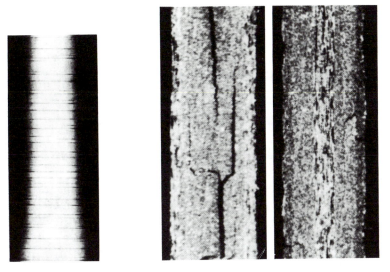

Fig. 12.22   **Fig. 12.23 a, b**

Fig. 12.22. Delaminations, as observed by contrast-enhanced X-ray radiography in a $[0°/\pm45°/90°]_s$ carbon/epoxy laminate upon fatigue testing. (From Ref. 49, used with permission.)

Fig. 12.23. Comparison of crack patterns at the free edges of a carbon/epoxy laminate: **a** $[0°/\pm45°/90°]_s$, **b** $[90°/\pm45°/0°]_s$. Note the greater severity of edge delaminations in **a** than in **b**. (From Ref. 49, used with permission.)

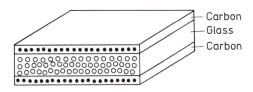

Carbon
Glass
Carbon

Fig. 12.24. Schematic of a hybrid laminate composite containing carbon and glass fibers

12.7.1 Hybrid Composite Systems

Yet another degree of flexibility in fiber composites is obtained by making what are called hybrid composites wherein one uses more than one type of fiber; see Fig. 12.24. Cost–performance effectiveness can be increased by judiciously using different reinforcement types and selectively placing them to get the highest strength in highly stressed locations and directions. For example, in a hybrid composite laminate the cost can be minimized by reducing the carbon fiber content, while the performance is maximized by "optimal placement and orientation of the fiber." Figure 12.25 shows the increases in specific flexural modulus (curve A) and specific tensile modulus (curve B) with weight % of carbon fiber in a hybrid composite [52]. Figure 12.26 shows the changes in flexural fatigue behavior as we go from 100% unidirectional carbon fibers in polyester (curve A) to unidirectional carbon faces over chopped glass core (curve B) to 100% chopped glass in polyester resin (curve C) [52].

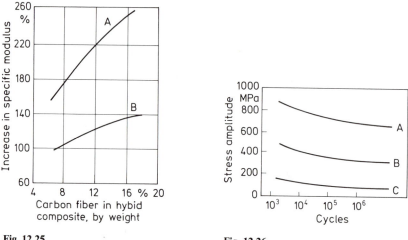

Fig. 12.25

Fig. 12.26

Fig. 12.25. Change in specific flexural modulus and specific tensile modulus with weight % carbon fibers. (From J.P. Riggs, Encyclopedia of Polymer Science & Engineering, 2nd. ed., vol. 2, copyright © 1985, John Wiley & Sons, New York. Reprinted by permission of John Wiley & Sons, Inc.) A = specific flexural modulus; B = specific tensile modulus

Fig. 12.26. Flexural fatigue behavior of 100% carbon fiber composite (curve A), carbon fiber facings with chopped glass fiber core (curve B), and 100% chopped glass fiber composite (curve C). (From J.P. Riggs, Encyclopedia of Polymer Science & Engineering, 2nd. ed., vol. 2, copyright © 1985, John Wiley & Sons, New York. Reprinted by permission of John Wiley & Sons, Inc.)

   An interesting hybrid composite system called ARALL, an acronym for aramid aluminum laminates, has been developed [53–54]. Initially developed at the Delft University of Technology in the Netherlands, the commercialization of ARALL was started by Alcoa and 3M Co. in the mid-eighties. ARALL consists of alternating layers of aluminum sheets bonded by adhesive containing aramid fibers. It is claimed that this unusual combination gives ARALL high strength, excellent fatigue and fracture resistance, combined with the advantages of metal construction such as formability, machinability, toughness, and impact resistance, etc. It can be formed, punched, riveted, or bolted like a metal. Potential applications for ARALL include "tension dominated fatigue and fracture critical structures" such as aircraft fuselage, lower wing and tail skins. In these structures, the use of ARALL will result in 15–30% weight savings over conventional construction.

References

1   K.K. Chawla and M. Metzger, *J. Mater. Sci.*, **7**, 34 (1972).
2   K.K. Chawla, *Metallography*, **6**, 155 (1973).
3   K.K. Chawla, *Philos. Mag.*, **28**, 401 (1973).
4   R.J. Arsenault and R.M. Fisher, *Scripta Met.*, **17**, 67 (1983).
5   K.K. Chawla, J. Singh, and J.M. Rigsbee, *Metallography*, **19**, 119 (1986).
6   M. Vogelman, R.J. Arsenault, and R.M. Fisher, *Met. Trans. A*, **17A**, 379 (1986).

7   L.J. Ebert and J.D. Gadd, in *Fiber Composite Materials*, ASM, Metals Park, OH, 1965, p. 89.

8   A. Kelly and H. Lilholt, *Philos. Mag.*, **20**, 311 (1969).

9   A. Kelly and G.J. Davies, *Metallurgical Rev.*, **10**, 1 (1965).

10 B.W. Rosen, in *Fiber Composite Materials*, American Society for Metals, Metals Park, OH, 1965, p. 58.

11 J. R. Lager and R.R. June, *J. Composite Mater.*, **3**, 48 (1969).

12 N.L. Hancox, *J. Mater. Sci.*, **10**, 234 (1975).

13 M.R. Piggott and B. Harris, *J. Mater. Sci.*, **15**, 2523 (1980).

14 M.R. Piggott, in *Developments in Reinforced Plastics — 4*, Elsevier Applied Science Publishers, London, 1984, p. 131.

15 D.K. Hale and A. Kelly, *Ann. Rev. Mater. Sci.*, **2**, 405 (1972).

16 J.O. Outwater and M.C. Murphy, in *Proceedings of the 24th SPI/RP Conference*, paper 11-6, Society of Plastics Industry, New York, 1969.

17 A. Kelly, *Proc. R. Soc. London*, **A319**, 95 (1970).

18 A.H. Cottrell, *Proc. R. Soc.*, **282A**, 2 (1964).

19 G.A. Cooper and A. Kelly, *J. Mech. Phys. Solids*, **15**, 279 (1967).

20 A. Kelly, in *The Properties of Fibre Composites*, IPS Science & Technology Press, Guildford, Surrey, U.K., 1971, p. 5.

21 G.A. Cooper, *J. Mater. Sci.*, **5**, 645 (1970).

22 J. Cook and J.E. Gordon, *Proc. R. Soc. London*, **A228**, 508 (1964).

23 H. Saghizadeh and C.K.H. Dharan, American Society of Mechanical Engineering, paper #85-WA/Mats-15, presented at the Winter Annual Meeting, Miami Beach, FL, Nov. 17–21, 1985.

24 B.D. Coleman, *J. Mech. Phys. Solids*, **7**, 60 (1958).

25 B.W. Rosen, in *Fiber Composite Materials*, American Society for Metals, Metals Park, OH, 1965, p.37.

26 B.W. Rosen, in *Mechanics of Composite Materials: Recent Advances*, Pergamon Press, Oxford, 1983, p. 105.

27 H.E. Daniels, *Proc. R. Soc.*, **A183**, 405 (1945).

28 C. Zweben and B.W. Rosen, *J. Mech. Phys. Solids*, **18**, 189 (1970).

29 R.E. Rowlands, in *Failure Mechanics of Composites*, vol. 3 of the series Handbook of Composites, North-Holland, Amsterdam, 1985, p. 71.

30 S.W. Tsai and E.M. Wu, *J. Composite Mater.*, **5**, 58 (1971).

31 S.W. Tsai and H.T. Hahn, *Introduction to Composite Materials*, Technomic, Westport, CT, 1980.

32 O. Hoffman, *J. Composite Mater.*, **1**, 200 (1967).

33 S.C. Cowin, *J. Appl. Mech.*, **46**, 832 (1979).

34 R.B. Pipes and B.W. Cole, *J. Compsite Mater.*, **7**, 246 (1973).

35 H.T. Hahn and L. Lorenzo, in *Advances in Fracture Research*, ICF6, New Delhi, India, Pergamon Press, Oxford, 1984, vol. 1, p. 549.

36 L.J. Broutman and S. Sahu, in *Proceedings of the 24th Annual Technical Conference*, Society of Plastics Industry, New York, 1969, Sect. 11-D, p. 1.

37 H.T. Hahn and R.Y. Kim, *J. Composite Mater.*, **10**, 156 (1976).

38 A.L. Highsmith and K.L. Reifsnider, in *Damage in Composite Materials*, ASTM STP 775, American Society of Testing & Materials, Philadelphia, 1982, p. 103.

39 N. Laws, G.J. Dvorak, and M. Hejazi, *Mech. Mater.*, **2**, 123 (1983).

40 R. Talreja, *Fatigue of Composite Materials*, Solid Mechanics Department, Technical University of Denmark, Lyngby, Denmark, 1985.

41 K.L. Reifsnider, E.G. Henneke, W.W. Stinchcomb, and J.C. Duke, in *Mechanics of Composite Materials*, Pergamon Press, New York, 1983, p. 399.

42 T.K. O'Brien and K.L. Reifsnider, *J. Composite Mater.*, **15**, 55 (1981).

43 R. Talreja, in *Advances in Composite Materials*, ICCM/3, Pergamon Press, Oxford, vol. 2, p. 1732.

44 S.L. Ogin, P.A. Smith, and P.W.R. Beaumont, *Composites Sci. Tech.*, **22**, 23 (1985).

45 A. Poursartip, M.F. Ashby, and P.W.R. Beaumont, in *Proceedings of the 3rd Risø International Symposium on Metallurgy & Materials Science*, Roskilde, Denmark, 1982, p. 279.

46 H.S. Kliger, *Machine Des.*, **51**, 150 (Dec. 6, 1979).

47 W.T. Fujimoto and B.R. Noton, in *Proceedings of the 6th St. Louis Symposium on Composite Material Engineering & Design*, American Society for Metals, Metals Park, OH, 1973, p. 335.

48   D.P. Seraphim, D.E. Barr, W.T. Chen, and G.P. Schmitt, in *Advanced Thermoset Composites*, Van Nostrand Reinhold, New York, 1986, p. 110.
49   H. Bergmann, in *Carbon Fibre and Their Composites*, Springer-Verlag, Berlin, 1985, p. 184.
50   P.L.N. Murthy and C.C. Chamis, *J. Composite Tech. Res.*, **8**, 8 (1986).
51   P.L.N. Murthy and C.C. Chamis, *ICAN: Integrated Composite Analyzer Users and Programmers' Manual*, NASA TP-2515, 1985.
52   J.P. Riggs, in *Encyclopedia of Polymer Science & Engineering*, 2nd ed., vol. 2, John Wiley & Sons, New York, 1985, p. 640.
53   L.B. Vogelsang and J.W. Gunnik, *Delft University of Technology Report LR-400*, Delft, The Netherlands, Aug., 1983.
54   L.N. Mueller, J.L. Prohaska, and J.W. Davis, "*ARALL: Introduction of a new composite material*", paper presented at the AIAA Aerospace Eng. Conf. & Show, Los Angeles, CA, Feb. 1985.

Suggested Reading

L.J. Broutman (ed.), *Fracture and Fatigue*, Academic Press, New York, 1974.
C.C. Chamis (ed.), *Structural Design and Analysis, Parts I and II*, Academic Press, New York, 1974.
B. Harris, *Fibre Reinforced Composites*, Institute of Metals, London, 1985.
W.S. Johnson (ed.), *Delamination and Debonding of Materials*, ASTM STP 876, American Society of Testing and Materials, Philadelphia, 1985.
G. Lubin (ed.), *Handbook of Composites*, Van Nostrand Reinhold, New York, 1982.
E. Scala, *Composite Materials for Combined Function*, Hayden Book Co., Rochelle Park, NJ, 1978.
M.M. Schwartz, *Composite Materials Handbook*, McGraw-Hill, New York, 1984.
R. Talreja, *Fatigue of Composite Materials*, Department of Solid Mechanics, Technical University of Denmark, Lyngby, Denmark, 1985.
S.W. Tsai, *Composites Design 1986*, Think Composites, Dayton, OH, 1986.

Appendices

Appendix A: Matrices

Definition

A matrix is an array of numbers subject to certain operations. For example,

$$\begin{bmatrix} 3 & 2 & 9 & 7 \\ 0 & -63 & 49 & 5 \\ 1 & 5 & 4 & 8 \end{bmatrix} \quad \text{3 rows, 4 columns}$$

$$\begin{bmatrix} a_{11} & a_{12} & \cdots & a_{1n} \\ a_{21} & a_{22} & \cdots & a_{2n} \\ \vdots & \vdots & \cdots & \vdots \\ a_{m1} & a_{m2} & \cdots & a_{mn} \end{bmatrix} \quad \text{m rows, n columns}$$

The a_{ij} are called the elements of the matrix.

Notation

$$A = [A] = (a_{ij}) = [a_{ij}]_{m,n}$$

Null Matrix

All the elements are zero, that is,

$$[N] = [0]$$

Any matrix multiplied by a null matrix results in a null matrix. Thus,

$$[A][N] = [N]$$

Square Matrix

The number of rows, called the order of the matrix, is equal to the number of columns. For example,

$$[A] = \begin{bmatrix} a_{11} & a_{12} \\ a_{21} & a_{22} \end{bmatrix}$$

is a square matrix of order 2.

Determinant

Associated with every square matrix $[A]$, there is a numerical value called its determinant, $|A|$.

$$|A| = a_{ij}(\mathrm{cof}\, a_{ij})$$

where $\mathrm{cof}\, a_{ij} = (-1)^{i+j}|M_{ij}|$. $\mathrm{cof}\, a_{ij}$ is the cofactor of a_{ij} and M_{ij} is the minor of a_{ij}, obtained by eliminating the row and the column of a_{ij}.

Consider the matrix

$$[A] = \begin{bmatrix} a_{11} & a_{12} \\ a_{21} & a_{22} \end{bmatrix}$$

Determinant $|A| = \begin{vmatrix} a_{11} & a_{12} \\ a_{21} & a_{22} \end{vmatrix}$

$$\mathrm{cof}\, a_{11} = (-1)^2|M_{11}| = M_{11} = a_{22}$$
$$\mathrm{cof}\, a_{12} = (-1)^3|M_{12}| = -M_{12} = -a_{21}$$
$$\mathrm{cof}\, a_{21} = (-1)^3|M_{21}| = -M_{21} = -a_{12}$$
$$\mathrm{cof}\, a_{22} = (-1)^4|M_{22}| = M_{22} = a_{11}$$

$$|A| = \sum_{i=1}^{n} a_{ij}(\mathrm{cof}\, a_{ij}) = a_{11}a_{22} + a_{12}(-a_{21})$$

$$= a_{11}a_{22} - a_{12}a_{21}$$

Another example is

$$[A] = \begin{bmatrix} 1 & 2 & -1 \\ 3 & 4 & 0 \\ 0 & 3 & 2 \end{bmatrix}$$

$$\mathrm{cof}\, a_{11} = M_{11} = \begin{vmatrix} 4 & 0 \\ 3 & 2 \end{vmatrix} = 8 - 0 = 8$$

$$\mathrm{cof}\, a_{12} = -M_{12} = \begin{vmatrix} 3 & 0 \\ 0 & 2 \end{vmatrix} = -6 + 0 = -6$$

$$\mathrm{cof}\, a_{13} = M_{13} = \begin{vmatrix} 3 & 4 \\ 0 & 3 \end{vmatrix} = 9 - 0 = 9$$

$$|A| = 1 \times 8 + 2 \times (-6) - 1 \times 9 = 8 - 12 - 9 = -13$$

Diagonal Matrix

A square matrix whose nondiagonal elements are zero is called a diagonal matrix.

$$a_{ij} = 0 \text{ for } i \neq j$$

For example,

$$\begin{bmatrix} 3 & 0 & 0 \\ 0 & 2 & 0 \\ 0 & 0 & 1 \end{bmatrix}$$

Identity Matrix

A diagonal matrix whose elements are unity is called an identity matrix.

$$[I] = \begin{bmatrix} 1 & 0 & 0 \\ 0 & 1 & 0 \\ 0 & 0 & 1 \end{bmatrix}$$

$$[I][A] = [A]$$

Column or Row Matrix

All the elements in a column or row matrix are arranged in the form of a column or row. For example,

$$\text{column matrix} \quad \begin{bmatrix} a_1 \\ a_2 \\ \cdot \\ \cdot \\ \cdot \\ a_n \end{bmatrix}$$

$$\text{row matrix} \quad [b_1 \quad b_2 \quad \cdots \quad b_n]$$

Transpose Matrix

The substitution of columns by rows of the original matrix produces a transpose matrix. The transpose of a row matrix is a column matrix and vice versa. For example,

$$[A] = \begin{bmatrix} 0 & -5 & 1 \\ 8 & 4 & 2 \end{bmatrix} \quad [A]^T = \begin{bmatrix} 0 & 8 \\ -5 & 4 \\ 1 & 2 \end{bmatrix}$$

$$[a] = [2 \quad 3 \quad -7] \quad [a]^T = \begin{bmatrix} 2 \\ 3 \\ -7 \end{bmatrix}$$

$$[AB]^T = [B]^T[A]^T$$

Symmetric Matrix

A square matrix is symmetric if it is equal to its transpose, that is,

$$A^T = A$$

$$a_{ij} = a_{ij} \quad \text{for all } i \text{ and } j$$

Antisymmetric or Skew-Symmetric Matrix

$$A^T = -A$$

$$a_{ij} = -a_{ji} \quad \text{for all } i \text{ and } j$$

For $i = j$, $a_{ii} = -a_{ii}$, that is the diagonal elements of a skew-symmetric matrix are zero.

Singular Matrix

A square matrix whose determinant is zero is called a singular matrix.

Inverse Matrix

$[A]^{-1}$ is the inverse matrix of $[A]$ such that

$$[A][A]^{-1} = [I] = [A]^{-1}[A]$$

$$[A]^{-1} = \frac{1}{|A|}[\text{cof } A]^T$$

We now give some examples of inverse matrix determination.

Example 1

$$A = \begin{bmatrix} a_{11} & a_{12} \\ a_{21} & a_{22} \end{bmatrix}$$

$$A^{-1} = \begin{bmatrix} a_{22} & -a_{12} \\ -a_{21} & a_{11} \end{bmatrix} \frac{1}{a_{11}a_{22} - a_{12}a_{21}}$$

Example 2

$$A = \begin{bmatrix} a_{11} & 0 & \cdots & 0 \\ 0 & a_{22} & \cdots & 0 \\ \vdots & \vdots & \cdots & \vdots \\ 0 & 0 & \cdots & a_{nn} \end{bmatrix}$$

$$A^{-1} = \begin{bmatrix} 1/a_{11} & 0 & \cdots & 0 \\ 0 & 1/a_{22} & \cdots & 0 \\ \vdots & \vdots & \cdots & \\ 0 & 0 & \cdots & 1/a_{nn} \end{bmatrix}$$

Example 3

$$A = \begin{bmatrix} \cos\theta & -\sin\theta \\ \sin\theta & \cos\theta \end{bmatrix}$$

$$|A| = \cos^2\theta + \sin^2\theta = 1$$

$$A^{-1} = \begin{bmatrix} \cos\theta & \sin\theta \\ -\sin\theta & \cos\theta \end{bmatrix}$$

Example 4

$$T = \begin{bmatrix} m^2 & n^2 & 2mn \\ n^2 & m^2 & -2mn \\ -mn & mn & m^2-n^2 \end{bmatrix} \qquad \begin{aligned} m &= \cos\theta \\ n &= \sin\theta \end{aligned}$$

This is a very important matrix called the transformation matrix.

$$|A| = 1$$

$$T^{-1} = \begin{bmatrix} m^2 & n^2 & -2mn \\ n^2 & m^2 & 2mn \\ mn & -mn & m^2-n^2 \end{bmatrix}$$

Note that T^{-1} is obtained by replacing θ by $-\theta$ in T.

Some Basic Matrix Operations

The equality $[A] = [B]$ is valid if and only if $a_{ij} = b_{ij}$.

Addition and Subtraction

To perform the operations of addition or subtraction the matrices must be compatible, that is, they must have the same dimension:

$$[A] + [B] + [C] = [D]$$

$$a_{ij} + b_{ij} = c_{ij} = d_{ij}$$

Similarly,

$$[E] = [A] - [B]$$

$$e_{ij} = a_{ij} - b_{ij}$$

Multiplication

To perform the operation of multiplication, the number of columns of one matrix must be equal to the number of rows of the other matrix:

$$[A]_{m \times n}[B]_{n \times r} = [C]_{m \times r}$$

$$c_{ij} = \sum_{k=1}^{n} a_{ik}b_{kj}$$

$$\begin{bmatrix} 3 & 5 \\ 9 & 8 \end{bmatrix}\begin{bmatrix} 1 & 0 \\ 4 & 3 \end{bmatrix} = \begin{bmatrix} 3 \times 1 + 5 \times 4 & 3 \times 0 + 5 \times 3 \\ 9 \times 1 + 8 \times 4 & 9 \times 0 + 8 \times 3 \end{bmatrix} = \begin{bmatrix} 23 & 15 \\ 41 & 24 \end{bmatrix}$$

Matrix multiplication is associative and distributive but not commutative. Thus,

$$[AB][C] = [A][BC]$$

$$[A][B + C] = [A][B] + [A][C]$$

$$[A][B] \neq [B][A]$$

The order of the matrix is very important in matrix multiplication.

Appendix B: Fiber Packing in Unidirectional Composites

Geometrical Considerations

Consider a composite containing uniaxially aligned fibers. Assume for the sake of simplicity that the fibers have the same cross-sectional form and area. Then for any uniaxial arrangement of fibers, we can relate the fiber volume fraction V_f, the fiber radius r, and the center-to-center spacing of fibers R as

$$V_f = \alpha \left(\frac{r}{R} \right)^2 \tag{1}$$

where α is a constant that depends on the geometry of the arrangement of fibers.

Let S be the distance of closest approach between the fiber surfaces. Then

$$S = 2(R - r) \tag{2}$$

or

$$S = 2 \left[\left(\frac{\alpha}{V_f} \right)^{1/2} r - r \right] = 2r \left[\left(\frac{\alpha}{V_f} \right)^{1/2} - 1 \right]$$

Thus, the separation between two fibers is less than a fiber diameter when

$$\left[\left(\frac{\alpha}{V_f} \right)^{1/2} - 1 \right] < 1$$

or

$$\left(\frac{\alpha}{V_f} \right)^{1/2} < 2$$

or

$$\frac{\alpha}{V_f} < 4$$

or

$$V_f > \frac{\alpha}{4} .$$

The densest fiber packing $V_{f_{max}}$ corresponds to touching fibers, that is,

$$\left[\left(\frac{\alpha}{V_{f_{max}}} \right)^{1/2} - 1 \right] = 0$$

or

$$V_{f_{max}} = \alpha$$

Table B.1 shows the values of α, $\alpha/4$, and $V_{f_{max}}$ for hexagonal and square arrangements of fibers.

Table B.1. Geometrical fiber packing parameters

Arrangement	α	$\alpha/4$	$V_{f_{max}}$
Hexagonal	$\dfrac{\pi}{2\sqrt{3}} = 0.912$	0.228	0.912
Square	$\dfrac{\pi}{4} = 0.785$	0.196	0.785

Appendix C: Some Important Units and Conversion Factors

Stress (or Pressure)

1 dyn $= 10^5$ newton (N)
$1 \, \text{N m}^{-2} = 10 \, \text{dyn cm}^{-2} = 1$ pascal (Pa)
1 bar $= 10^5 \, \text{N m}^{-2} = 10^5$ Pa
1 hectobar $= 100$ bars $= 10^8 \, \text{cm}^{-2}$
1 kilobar $= 10^8 \, \text{N m}^{-2} = 10^9 \, \text{dyn cm}^{-2}$
1 mm Hg $= 1$ torr $= 133.322$ Pa $= 133.322 \, \text{N m}^{-2}$
$1 \, \text{kgf mm}^{-2} = 9806.65 \, \text{kN m}^{-2} = 9806.65$ kPa $= 100$ atmospheres
$1 \, \text{kgf cm}^{-2} = 98.0665$ kPa $= 1$ atmosphere
$1 \, \text{lb in}^{-2} = 6.89476 \, \text{kN m}^{-2} = 6.89476$ kPa
$1 \, \text{kgf cm}^{-2} = 14.2233 \, \text{lb in}^{-2}$
10^6 psi $= 10^6 \, \text{lb in}^{-2} = 6.89476 \, \text{GN m}^{-2} = 6.89476$ GPa
1 GPa $= 145\,000$ psi

Density

$1 \, \text{g cm}^{-3} = 62.4280 \, \text{lb ft}^{-3} = 0.0361 \, \text{lb in}^{-3}$
$1 \, \text{lb in}^{-3} = 27.68 \, \text{g cm}^{-3}$
$1 \, \text{g cm}^{-3} = 10^3 \, \text{kg m}^{-3}$

Viscosity

1 poise $= 0.1$ Pa s $= 0.1 \, \text{Nm}^{-2}$ s
$1 \, \text{GN m}^{-2}$ s $= 10^{10}$ poise

Energy per Unit Area

$1 \, \text{erg cm}^{-2} = 1 \, \text{mJ m}^{-2} = 10^{-3} \, \text{J m}^{-2}$
$10^8 \, \text{erg cm}^{-2} = 47.68 \, \text{ft lb in}^{-2} = 572.16$ psi in

Fracture Toughness

1 psi $\text{in}^{1/2} = 1 \, \text{lbf in}^{-3/2} = 1.11 \, \text{kN m}^{-3/2} = 1.11 \, \text{kPa m}^{-1/2}$
1 ksi $\text{in}^{1/2} = 1.11 \, \text{MPa m}^{-1/2}$
$1 \, \text{MPa m}^{-1/2} = 0.90$ ksi $\text{in}^{1/2}$
$1 \, \text{kgf mm}^{-2} \, \text{mm}^{1/2} = 3.16 \times 10^4 \, \text{N m}^{-2} \, \text{m}^{1/2}$

Problems

Chapter 1

1.1. Describe the structure and properties of some fiber composites that occur in nature.

1.2. Many ceramic based composite materials are used in electronics industry. Describe some of these electroceramic composites.

1.3. Describe the use of composite materials in the Voyager airplane that circled the globe for the first time without refueling in Dec. 1986.

Chapter 2

2.1. Nonwoven fibrous mats can be formed through entanglement and/or fibers bonded in the form of webs or yarns by chemical or mechanical means. What are the advantages and disadvantages of such nonwovens over similar woven mats?

2.2. The compressive strength of Kevlar fiber is about one-eighth of its tensile stress. Estimate the smallest diameter of a rod on which the Kevlar fiber can be wound without causing kinks etc. on its compression side.

2.3. What is asbestos fiber and why is it considered to be a health hazard?

2.4. Glass fibers are complex mixtures of silicates and borosilicates containing mixed sodium, potassium, calcium, magnesium, and other oxides. Such a glass fiber can be regarded as an inorganic polymeric fiber. Do you think you can provide the chemical structure of such a polymer chain?

2.5. A special kind of glass fiber is used as a medium for the transmission of light signals. Discuss the specific requirements for such an optical fiber.

2.6. What are the major sources of fiber stress in any optical cable design?

2.7. Describe the problems involved in mechanical testing of whiskers.

2.8. Several types of Kevlar aramid fibers are available commercially. Draw the stress strain curves of Kevlar 49 and Kevlar 29. Describe how much of the strain is elastic (linear or nonlinear). What microstructural processes occur during their deformation?

2.9. Describe the structural differences between Kevlar and Nomex (both aramids) that explain their different mechanical characteristics.

2.10. Explain why when the Kevlar fiber is fractured longitudinal splitting (micro-fibrillation) is characteristically observed.

Chapter 3

3.1. Ductility, the ability to deform plastically in response to stresses, is more of a characteristic of metals than it is of ceramics or polymers. Why?

3.2. Ceramic materials generally have some residual porosity. How does the presence of porosity affect the elastic constants of ceramic materials? How does it affect the fracture energy of ceramics?

3.3. Explain why is it difficult to compare the stress-strain behavior of polymers (particularly thermoplastics) with that of metals.

3.4. The mechanical deformation of a polymer can be represented by an elastic spring and a dashpot in parallel (Voigt model). For such a model we can write for stress

$$\sigma = \sigma_{el} + \sigma_{visc} = E\varepsilon + \eta\frac{d\varepsilon}{dt}$$

where E is the Young's modulus, ε is the strain, η is the coefficient of viscoelasticity, and t is the time. Show that

$$\varepsilon = \frac{\sigma}{E}\left[1 - \exp\left(-\frac{E}{\eta}\right)t\right].$$

3.5. What is the effect of the degree of crystallinity on fatigue resistance of polymers?

3.6. Discuss the importance of thermal effects (internal heating) on fatigue of polymers.

3.7. Glass-ceramics combine the generally superior electrical and mechanical properties of crystalline ceramics with the processing ease of glasses. Give a typical thermal cycle involving the various stages for producing a glass-ceramic.

Chapter 4

4.1. Describe some techniques, using SEM and TEM, for measuring interfacial energies in metal/metal and metal/ceramic systems.

4.2. In order to study the interfacial reactions between the fiber and matrix, oftentimes one uses very high temperatures in order to reduce the time necessary for the experiment. What are the objections to such accelerated tests?

4.3. What are the objections to the use of short beam shear test to measure the interlaminar shear strength (ILSS)?

Chapter 5

5.1. Why are prepregs so important in polymer matrix composites? What are their advantages? Describe the different types of prepregs.

5.2. Randomly distributed short fibers should result in more or less isotropic properties in an injection molded composite. But this is generally not true. Why? What are the other limitations of injection molding process?

Chapter 6

6.1. Discuss the advantages and disadvantages of casting vis a vis other methods of fabricating metal matrix composites.

6.2. Silicon carbide (0.1 μm thick) coated boron fiber was used to reinforce a metallic matrix. The SiC coating serves as a diffusion barrier coating. Estimate the time for dissolution of this coating at 700 K if the diffusion coefficient at 700 K is 10^{-16} m^2 s^{-1}.

6.3. The metallic matrix will generally undergo constrained plastic flow in the presence of a moderately high volume fraction of high modulus fibers. Draw schematically the stress-strain curves of a constrained metal matrix (i.e. insitu behavior) and an unconstrained metal (i.e. 100% matrix metal). Explain the difference.

6.4. A spray forming technique, called the Osprey process, is very promising for producing solid preform billets, tubes, sheets directly from the melt. The charge is induction melted and the liquid metal droplets are sprayed on a substrate. Maneuvering of the substrate in different ways can result in a variety of shapes. Comment on the use of this technique for producing ceramic particle reinforced metal matrix composites.

6.5. Discuss the problem of thermal stability of unidirectionally solidified eutectic (insitu) metallic composites.

Chapter 7

7.1. What are the sources of fiber degradation during processing of ceramic matrix composites?

7.2. Describe the advantages of using sol-gel and polymer pyrolysis techniques to process the ceramic matrix in CMCs.

7.3. Explain how a carbon fiber reinforced glass-ceramic composite can be obtained with an almost zero in-plane coefficient of thermal expansion.

Chapter 8

8.1. In view of the highly inert and non-leachable nature of carbon fibers, what waste disposal practices can you recommend for carbon fibers and their composites?

8.2. List the various matrix resins that are commonly used with carbon fibers.

8.3. Distinguish between interphase and interface.

Chapter 9

9.1. There are many known superconducting A15 compounds. Of these Nb_3Al, Nb_3Ga, and Nb_3Ge have higher values of T_c and Hc_2 than do Nb_3Sn and V_3Ga. How then does one explain the fact that only Nb_3Sn and to a lesser extent V_3Ga are available commercially?

9.2. It is believed that grain boundaries are the imperfections responsible for the flux-pinning in high-J_c materials like Nb_3Sn and V_3Ga. How does J_c vary with grain size?

9.3. What is the effect of any excess unreacted bronze leftover in the manufacture of Nb_3Sn superconductor composite via the bronze route?

9.4. Examine the Nb-Sn phase diagram. At what temperature does the A15 compound (Nb_3Sn) become unstable? Nb_3Sn is formed by solid state diffusion in Nb/Cu-Sn composites at 700°C or below. Is this in accord with information from the phase diagram? Explain.

9.5. Do you think it is important to study the effect of irradiation on superconducting materials? Why?

9.6. In the high magnetic field coils of large dimensions, rather large tensile and compressive loads can be encountared during energizing and deenergizing. Discuss the effects of cyclic stress on the superconducting coil materials.

9.7. Describe some experimental methods of measuring void content in composites. Give the limitations of each method.

9.8. Superconducting composites in large magnets can be subjected to high mechanical loads. Describe the sources of these.

Chapter 10

10.1. Consider a 40% V_f SiC whisker reinforced aluminum composite. $E_f = 400$ GPa, $E_m = 70$ GPa, and $(l/d) = 20$. Compute the longitudinal elastic modulus of this composite if all the whiskers are aligned in the longitudinal direction. Use Halpin-Tsai equations. Take $\xi = 2(l/d)$.

10.2. A composite has 40% V_f of a 150 μm diameter fiber. The fiber strength is 2.5 GPa, the matrix strength is 75 MPa, while the fiber/matrix interfacial strength is 50 MPa. Assuming a linear build up of stress from the two ends of a fiber, estimate the composite strength for (a) 200 mm long fibers and (b) 3 mm long fibers.

10.3. Derive the load transfer expression (Eq. 10.51) using the boundary conditions. Show that average tensile stress in the fiber is given by Eq. (10.52).

10.4. Consider a fiber composite system in which the fiber has an aspect ratio of 1000. Estimate the minimum interfacial shear strength τ_i, as a percentage of the tensile stress in fiber, σ_f, which is necessary to avoid interface failure in the composite.

10.5. Show that as $\xi \to 0$, the Halpin-Tsai equations reduce to

$$1/p = V_m/p_m + V_f/p_f$$

while as $\xi \to \infty$, they reduce to

$$p = V_m p_m + V_f p_f .$$

10.6. Consider an alumina (FP) fiber reinforced magnesium composite. Calculate the composite stress at the matrix yield strain. The matrix yield stress is 180 MPa, $E_m = 70$ GPa, and $v = 0.3$. Take $V_m = 50\%$

10.7. Estimate the aspect ratio and the critical aspect ratio for aligned SiC whiskers (5 μm diameter and 2 mm long) in an aluminum alloy matrix. Assume that the matrix alloy does not show any work hardening.

10.8. Alumina whiskers (density $= 3.8$ g cm^{-3}) are incorporated in a resin matrix (density $= 1.3$ g cm^{-3}). What is the density of the composite? Take $V_f = 0.35$. What is the relative mass of the whiskers?

10.9. Consider a composite made of aligned, continuous boron fibers in an aluminum matrix. Compute the elastic moduli, parallel and transverse to the fibers. Take $V_f = 0.50$.

10.10. Fractographic observations on a fiber composite showed that the average fiber pullout length was 0.5 mm. If $V_{fu} = 1$ GPa and the fiber diameter is 100 μm, calculate the strength of the interface in shear.

10.11. Consider a tungsten/copper composite with following characteristics: fiber fracture strenght $= 3$ GPa, fiber diameter $= 200$ μm, and the matrix shear yield strength $= 80$ MPa. Estimate the critical fiber length which will make it possible that the maximum load bearing capacity of the fiber is utilized.

10.12. Carbon fibers ($V_f = 50\%$) and polyimide matrix have the following parameters:

$$E_f = 280 \text{ GPa} \qquad E_m = 276 \text{ MPa}$$

$$v_f = 0.2 \qquad v_m = 0.3$$

(a) Compute the elastic modulus in the fiber direction, E_{11}, and transverse to the fiber direction, E_{22}.
(b) Compute the Poisson ratios, v_{12} and v_{21}.

10.13. Copper or aluminum wires with steel cores are used for electrical power transmission. Consider a Cu/steel composite wire having the following data:

inner diameter $= 1$ mm

outer diameter $= 2$ mm

$$E_{Cu} = 150 \text{ GPa} \qquad \alpha_{Cu} = 16 \times 10^{-6} \text{ K}^{-1}$$

$$E_{steel} = 210 \text{ GPa} \qquad \alpha_{steel} = 11 \times 10^{-6} \text{ K}^{-1}$$

$$\sigma_{y Cu} = 100 \text{ MPa} \qquad \nu_{Cu} = \nu_{steel} = 0.3$$

$$\sigma_{y steel} = 200 \text{ MPa}$$

(a) The composite wire is loaded in tension. Which of the two components will yield plastically first? Why?

(b) Compute the tensile load that the wire will support before any plastic strain occurs.

(c) Compute the Young's modulus and coefficient of thermal expansion of the composite wire.

Chapter 11

11.1. An isotropic material is subjected to a uniaxial stress. Is the strain state also uniaxial? Write the stress and strain in matrix form.

11.2. For a symmetric laminated composite, we have, $\bar{Q}_{ij}(+z) = \bar{Q}_{ij}(-z)$, i.e., the moduli are even functions of thickness z. Starting from the definition $B_{ij}(= \int_{-h/2}^{h/2} \bar{Q}_{ij} z \, dz)$, split the integral and show that B_{ij} is identically zero for a symmetric laminate.

11.3. An orthotropic lamina has the following characteristics: $E_{11} = 210$ PGa, $E_{22} = 8$ GPa, $G_{12} = 5$ GPa, and $\nu_{12} = 0.3$. Consider a three-ply laminate made of such laminae arranged at $\theta = \pm 60°$. Compute the submatrices $[A]$, $[B]$, and $[D]$. Take the ply thickness to be 0.1 mm.

11.4. Enumerate the various phenomena which can cause microcracking in a fiber composite.

11.5. A thin lamina of a composite with fibers aligned at 45° to the lamina major axis is subjected to the following stress system along its geometric axes:

$$[\sigma_i] = \begin{bmatrix} \sigma_x \\ \sigma_y \\ \sigma_s \end{bmatrix} = \begin{bmatrix} 10 \\ 2 \\ 3 \end{bmatrix} \text{MPa} .$$

Compute the stress components along the material axes (i.e., σ_1, σ_2, and σ_6).

11.6. A two-ply laminate composite has the top and bottom ply orientations of 45° and 0° and thicknesses of 2 and 4 mm, respectively. The stiffness matrix for the 0° ply is

$$[Q_{ij}] = \begin{bmatrix} 20 & 1 & 0 \\ 1 & 3 & 0 \\ 0 & 0 & 1 \end{bmatrix} \text{GPa} .$$

Find the $[\bar{Q}_{ij}]_{45}$ and then compute the matrices $[A]$, $[B]$, and $[D]$ for this laminate.

11.7. A two-ply laminate composite is made of polycrystalline, isotropic aluminum and steel sheets, each 1 mm thick. The constitutive equations for the two sheets are:

$$\text{Al:} \begin{bmatrix} \sigma_1 \\ \sigma_2 \\ \sigma_6 \end{bmatrix} = \begin{bmatrix} 70 & 26 & 0 \\ 26 & 70 & 0 \\ 0 & 0 & 26 \end{bmatrix} \begin{bmatrix} \varepsilon_1 \\ \varepsilon_2 \\ \varepsilon_6 \end{bmatrix} \text{MPa}$$

$$\text{Steel:} \begin{bmatrix} \sigma_1 \\ \sigma_2 \\ \sigma_6 \end{bmatrix} = \begin{bmatrix} 210 & 60 & 0 \\ 60 & 210 & 0 \\ 0 & 0 & 78 \end{bmatrix} \begin{bmatrix} \varepsilon_1 \\ \varepsilon_2 \\ \varepsilon_6 \end{bmatrix} \text{MPa} .$$

Compute the matrices $[A]$, $[B]$ and $[D]$ for this laminate composite. Point out any salient features of this laminate.

Chapter 12

12.1. For a carbon/epoxy composite, the strength parameters are:

$$F_{11}(\text{GPa})^{-2} = 0.45$$
$$F_{22}(\text{GPa})^{-2} = 101$$
$$F_{12}(\text{GPa})^{-2} = -3.4$$
$$F_{66}(\text{GPa})^{-2} = 215$$
$$F_1(\text{GPa})^{-1} = 0$$
$$F_2(\text{GPa})^{-1} = 21 .$$

Compute the off axis uniaxial strengths of this composite for different θ and obtain a plot of σ_x vs. θ.

12.2. Acoustic emission can be used to monitor damage in carbon fiber/epoxy during fatigue. Under steady loading conditions the damage is controlled by fiber failure and one can describe the acoustic emission by

$$\frac{dN}{dt} = \frac{A}{(t+T)^n}$$

where N is the total number of emissions, t is the time, T is a time constant, and A is constant under steady loading conditions. Taking $n = 1$, show that $\log dt$ is a linear function of the accumulated counts. (Hint: see M. Fuwa, B. Harris, and A.R. Bunsell, *J. App. Phys.*, **8** (1975) 1460).

12.3. For a ceramic fiber with $\mu = 12\%$, show that $\beta \simeq 10$. Show also that if the fiber length is changed by an order of magnitude, the corresponding drop in the average strength is about 20 percent.

12.4. The average stress in a fiber ($E_f = 400$ GPa) is given by

$$\bar{\sigma}_f = \sigma_f[1 - (1 - \beta)l_c/l]$$

where β is the coefficient of load transfer from matrix to fiber, l_c is the critical fiber length, and l is the fiber length. The fiber behaves elastically and breaks at a strain of 0.1%. The matrix has a $\sigma_{mu} = 1$ MPa. Compute the V_{crit} for this system. Take $\beta = 0.5, \alpha = l/l_c = 10$, and the matrix stress corresponding to a strain of 0.1% equal to 0.5 MPa.

12.5. In a series of tests on boron fibers, it was found that $\mu = 10\%$. Compute the ratio $\bar{\sigma}_B/\bar{\sigma}$, where $\bar{\sigma}_B$ is the average strength of the fiber bundle and $\bar{\sigma}$ is the average strength of fibers tested individually.

12.6. List the possible fatigue crack initiating sites in fiber composites.

12.7. What factors you think will be important in the environmental effects on the fatigue behavior of fiber reinforced composites?

12.8. Estimate the work of fiber pullout in a 40% carbon fiber/epoxy composite. Given $\sigma_{fu} = 0.2$ GPa, $d = 8$ μm, and $\tau_i = 2$ MPa.

Author Index

Subject Index